Star Myths of the Greeks and Romans

Atlas Farnese

Star Myths
of the Greeks
and Romans

A Sourcebook

Containing *The Constellations*
of Pseudo-Eratosthenes
and the *Poetic Astronomy*
of Hyginus

**Translation and Commentary
by Theony Condos**

Phanes Press

Theony Condos holds a doctorate in classical studies from the University of Southern California. *Star Myths of the Greeks and Romans* is a revised and expanded version of her dissertation, *The* Katasterismoi *of the Pseudo-Eratosthenes: A Mythological Commentary and English Translation*.

© 1997 by Theony Condos.

Published by Phanes Press, PO Box 6114, Grand Rapids, MI 49516, U.S.A.

Phanes Press website: *www.phanes.com*

Library of Congress Cataloging-in-Publication Data

Star myths of the Greeks and Romans : a sourcebook containing the
    Constellations of Pseudo-Eratosthenes and the Poetic astronomy of
    Hyginus / translation and commentary by Theony Condos.
        p.   cm.
    Includes bibliographical references and index.
    ISBN 1-890482-92-7 (alk. paper)
    ISBN 1-890482-93-5 (pbk. : alk. paper)
    1. Constellations. 2. Astronomy, Ancient. 3. Mythology, Greek.
4. Mythology, Roman. 5. Eratostene. Constellations. 6. Hyginus.
Poetic astronomy. I. Eratostene. Constellations. II. Hyginus.
Poetic astronomy.
QB802.S83   1997
398.26--dc21   97-29207
                                                                   CIP

Printed on permanent, acid-free paper.
Printed in the United States of America.

Contents

6

CONTENTS

Γιά τούς γονείς μου

Preface

The present volume provides an English translation and commentary for two classical texts, the *Catasterismi* of Pseudo-Eratosthenes (first/second century C.E.), and Book 2 of the Latin work variously titled *Poeticon Astronomicon* or *De Astronomia*, which is attributed to Hyginus (first century B.C.E.). Together, these two texts offer a comprehensive picture of the myths associated by the Greeks and Romans with the constellations familiar to them. Those constellations were forty-eight in number by the time of Ptolemy (second century C.E.).

The translation of the *Catasterismi* follows the edition of Olivieri, which shows, separately, the consensus of readings of the five complete manuscripts and of manuscript R (Venetus Marcianus 444), a partial manuscript that stems from a different archetype than the complete manuscripts and contains some differences from them. Where manuscript R differs significantly from the consensus of readings in the other manuscripts, as for example, in the myth associated with the constellation Corona Borealis, the variant readings of R are identified by angle brackets (< . . . >) in the translation below. Lacunae in the text are identified by square brackets ([. . .]). The translation of the *De Astronomia* follows the edition of Viré.

Greek and Latin names are retained in form, but the spelling of Greek names is Latinized, e.g., Heracles and Hercules. A list of Greek names and their Latin counterparts is provided in Appendix 1.

The spot illustrations at the beginning of each chapter are taken from woodcuts in the first edition of Hyginus, *Poeticon Astronomicon* (Venice: Erhard Ratdolt, 1482). In most cases, the illustrations do not accurately depict the location or number of the stars described in the Greek and Latin texts translated below. A more accurate depiction of the constellations can be found in the star maps reproduced in

Appendix 3. The reader should also note that most star maps and pictures of the constellations show the constellations as they would appear from the Earth, i.e., from *inside* the celestial sphere, while many texts describe the figures as they would appear from *outside* the celestial sphere.

The proposed identification of stars by their modern designation in the translation of the *Catasterismi* is based on a comparison of the language used in the Greek text and in Ptolemy's *Almagest* to describe the location of a particular star. Where the descriptions are similar, e.g., "there is one star on the head," "the star on the head," identification is fairly straightforward. However, when the descriptions do not coincide, e.g., "there is one star on the head," "the northernmost of the three stars on the head," identification is more problematic. In such cases, the star of the greatest magnitude is proposed, followed by a question mark. When the *Catasterismi* describes a star for which there is no corresponding reference in Ptolemy, a question mark appears in the translation. Since Hyginus's descriptions of star locations rarely vary from those in the *Catasterismi*, an identification is proposed only in those instances when Hyginus's text varies significantly from the Greek. The proposed modern designation of stars listed in Ptolemy's *Almagest* was derived from two recent works: Paul Kunitzsch, *Der Almagest* (Wiesbaden: Otto Harrassowitz, 1974) and G. J. Toomer, *Ptolemy's Almagest* (New York: Springer-Verlag, 1984). It should be noted that the stars enumerated in the Greek and Latin texts do not always coincide with the total number indicated in those same texts.

When the star identified by a modern designation belongs to the constellation that is the subject of the Greek text, no constellation name is indicated. When, however, the star belongs to another constellation, a constellation name is noted, e.g., the stars in the "Claws of the Scorpion" belong not to the constellation Scorpio, but to the constellation Libra; thus, in the translation of the Greek text that treats Scorpio they are designated as α Lib, μ Lib, etc., following the standard abbreviations for the names of the constellations listed in Appendix 2.

Thanks are due to David Fideler for his interest in the subject, to Gibson Reaves for encouragement in a distant time and place, to Apostolos Athanassakis for assistance in interpreting a troublesome passage in the Greek text, to Brigitte Cazelles for a perceptive reading of the manuscript, and to Justin Chang for technical assistance. Responsibility for any remaining shortcomings rests, of course, with the author.

Introduction

The campaigns of Alexander the Great during the last half of the fourth century B.C.E. expanded the horizons of the Greek world by bringing Greeks and their culture into direct contact with the civilizations of the Ancient Near East. The ensuing juxtaposition of differing perspectives regarding the state, the gods, and the individual caused learned men throughout the Eastern Mediterranean to reassess their understanding of the world and human experience. An explosion of scientific inquiry and new knowledge resulted that was unequaled until modern times. By the third century B.C.E., Alexandria in Egypt had become one of the principal centers of literary and scientific studies in the Mediterranean world. At the center of this scholarly activity were two institutions: the famous Library, founded by Ptolemy I (311–283 B.C.E.), organized under Ptolemy II (283–246 B.C.E.), and maintained by their successors, and the lesser-known Museum, also established by Ptolemy I, as a residential center for research. The Museum's history is obscure—it appears to have been the victim of political tensions—while the Library, under a succession of learned head librarians including Zenodotus of Ephesus, Apollonius of Rhodes, and Eratosthenes of Cyrene, flourished for over 500 years, until its burning by Aurelian in 272 C.E.[1]

During the Hellenistic period (323–30 B.C.E.), intellectual activity in Alexandria produced significant results in two broad areas: science and literary scholarship. Guided by the Aristotelian model of scientific inquiry—that is, of gathering, classifying, and studying all available data before arriving at a conclusion—Alexandrian scientists achieved spectacular advances in fields such as astronomy, mathematics, medicine, geography, architecture, and urban planning.[2] In the literary sphere, Alexandrian scholars collected, catalogued, and systematically studied the Greek literature of the past, giving rise to the field of textual criticism. They prepared critical editions of Homer, drew up "canons" of authors by genre, including lyric, epic, tragedy, and comedy, and wrote commentaries on older literary works. Out of this intensive study of earlier Greek authors developed a literary aesthetic that became the uniquely Alexandrian contribution to Greek and Latin literature. Eschewing the universal themes of Greek epic and tragedy, numerous Alexandrian writers, the most famous of whom was Callimachus (c. 310–c. 240 B.C.E.), cultivated genres such as the epyllion (a kind of "mini-epic" focused on a single episode in the life of the protagonist), pastoral, hymn, and epigram—genres that were amenable to the treatment of more personal, sometimes even mundane, themes. These writers sought to impress their audience and each other with their craftsmanship: polished language, the felicitous turn of phrase, breadth of learning, and wit. The new aesthetic was reflected in the advice Callimachus claimed he had received from Apollo: "tread the paths the wagons do not go by; do not drive your chariot in the paths of others, nor on the broad road, but by untrodden paths, even if you drive a narrower way."[3] Finely wrought literary productions of moderate length were the order of the day and were eagerly received by an increasingly literate public. Callimachus described his audience as being "those who like the clear note of the cicada, not the noise of donkeys." Changing literary tastes were reflected in the great literary debate of the period—perhaps the original contest of the "ancients" and the "moderns"—which pitted Callimachus, as the champion and primary exponent of the emerging Alexandrian preference for narrowly focused, sophisticated, and highly polished literary productions, against Apollonius of Rhodes, himself the author of a long epic poem about the quest for the golden fleece, as defender of the older tradition of lengthy literary productions that treated universal themes, albeit with an Alexandrian infusion of romanticized sentiment.[4]

The presence of an avid reading public combined with the literary tastes and scientific proclivities of the age inspired a type of literary experiment that was peculiarly Alexandrian: the compilation and literary treatment of thematically related knowledge. Thus we encounter poems enumerating the antidotes to snake-bite, tracts on cooking, and poetic handbooks for amateur astronomers or fishermen.[5] To the extent these literary productions were intended to instruct their readers—and that intent was surely present—such poems and treatises were the most recent manifestation of a long tradition of didactic literature that included Hesiod and his Ancient Near Eastern antecedents.[6] The original contribution of Alexandria to the didactic tradition was to combine attention to literary form with originality of content, and, once again, to focus more narrowly on the human experience by offering the reader knowledge not as a guide for social interaction or for survival, but for its own sake. Thus, while Hesiod offered his reader advice on when to plough and when to sail, Alexandrian authors offered their readers a glimpse into a realm of arcane knowledge which, although entertaining, usually had little practical value. Later critics sometimes took a dim view of this type of literature. Strabo in the early first century C.E., for instance, took Eratosthenes soundly to task for having said that the purpose of poetry is to entertain rather than to instruct its reader.[7]

Perhaps related to this peculiarly Alexandrian didactic literature was an intense interest in aetiology, as manifested in the profusion of literary works dealing with origins, i.e., how various aspects of the physical world, customs, cults, or cities came to be.[8] Scientific learning, as represented by the didactic literature of the day coexisted happily with aetiological literature, which drew almost exclusively on myth and legend to achieve its purpose. Indeed, Eratosthenes of Cyrene embodied the two approaches to reality in his own writings.

Eratosthenes was about thirty years of age in 245 B.C.E., when he was summoned by Ptolemy III (246–221 B.C.E.) from Athens, where he was pursuing philosophical studies, to assume the post of head librarian of the Library at Alexandria, a post he held until his death, at age eighty, in 194 B.C.E. As head of the Library, and himself a scholar and poet, Eratosthenes was at the center of the intellectual ferment in Alexandria during the last half of the third century. His scholarly endeavors covered numerous fields: lexicography, chronology, geography, literary history, mathematics, and philosophy. Of his scholarship, we have little more than titles: he wrote a lengthy treatise on ancient comedy; his geographical works included *On the Measurement of the Earth*, in which he calculated the circumference of the Earth as well as distances between cities; his philosophical essays treated themes such as wealth and poverty and good and evil; he compiled a list of Olympic victors, and he wrote the *Chronographiae*, one of the first systematic chronologies.[9] Although a vigorous participant in the scholarly activity in his day, Eratosthenes appears to have been more a product than an architect of his age. His own contemporaries assigned to him the nickname "Beta," judging him to be only "second best" at any one his many pursuits.

As a pupil of Callimachus, Eratosthenes was squarely in the camp of the "moderns" in the literary debate between Callimachus and Apollonius. And, indeed, he appears to have enjoyed a good reputation as a poet. A few verses survive of his *Erigone*, an elegiac poem recounting how Dionysus introduced wine to mankind through Icarius, bringing about the tragic death of both Icarius and his daughter Erigone.[10] The literary critic Longinus (second/third century C.E.) refers to Eratosthenes's *Erigone* as a "flawless little poem."[11] Another of Eratosthenes's poems, the *Hermes*, consisted of about 1600 hexameter verses. The precise content of that poem is unclear; one surviving verse notes that "the planets possess the same harmony as the lyre."

Among Eratosthenes's prose works was the *Catasterismi*, a compilation of myths explaining the origin of the forty-eight constellations familiar to the Greeks of the Hellenistic era. The *Catasterismi* of Eratosthenes is not extant. What survives under this title is a collection of forty-four stories in all, forty-two explaining the origin of the various constellations, and two additional stories, one providing an account of the origin of the Milky Way, the other enumerating the names of the five planets. It is generally agreed that these forty-four stories constitute an epitome of the original work by Eratosthenes

compiled by an anonymous author labeled Pseudo-Eratosthenes
(hereafter Ps-Eratosthenes) in the first or second century C.E. The
question of whether the *Catasterismi* of Eratosthenes can be extricated
from the surviving epitome (hereafter *The Constellations*) occupied
many classical scholars in the late nineteenth and early twentieth
centuries—without great success.[12]

Eratosthenes was one of several Greek authors to devote an entire
literary work to the heavens. We know of three poetical astronomies,
now lost, by Cleostratus of Tenedos (sixth century B.C.E.), Sminthes
(fourth? century B.C.E.), and Alexander Aetolus (fourth/third centuries B.C.E.), respectively. We know also that the sole surviving poetical
astronomy, the *Phaenomena* of Aratus, was a versification of the prose
work of the same name by Eudoxus of Cnidus (fourth century B.C.E.).

The *Phaenomena* of Aratus of Soloi (315–250? B.C.E.), an older
contemporary of Eratosthenes, survives in its entirety; it is a poem of
some 1150 verses describing the relative positions of the constellations, with occasional reference to a myth associated with a particular
constellation. Its purpose was clearly descriptive, i.e., to lay out for the
reader the organization of the stars in the sky. The *Phaenomena* was a
singularly popular and influential work in antiquity, inspiring a score
of Greek imitations, Roman translations—including those by Cicero
(first century B.C.E.), Germanicus (first century C.E.), Avienus (fourth
century C.E.), and the eighth century C.E. author of the *Aratus
Latinus*—as well as lengthy commentaries in both Greek and Latin.
The reasons for the popularity of the *Phaenomena* are rendered elusive
by the passage of time.[13] They are, perhaps, related to Aratus's
espousal of Stoic ideas, or it may be, simply, that the *Phaenomena* was
a particularly useful guide to the stars, whose presence was more
strongly felt in a world without artificial light than it is in our own
enlightened age.

It is not clear to what extent the authors preceding Aratus included
in their works mythological explanations for the origin of the constellations. In the few instances when Aratus himself alludes to a myth, as
for example, in connection with the constellation Virgo, the myth

serves less as an aetion for the constellation than as support for the tenets of Stoic philosophy. Unless Aratus's sparse allusions to constellation myths are misleading, it may be safe to infer that the earlier works did not provide a mythological explanation for each constellation. It would appear, then, that while the works of Aratus and his predecessors described the location of constellation figures in the sky, it was Eratosthenes who first systematically assembled mythological material associated with each of the constellation figures. The only work of similar intent by a classical author is the *Poeticon Astronomicon* or *De Astronomia* (hereafter *Poetic Astronomy*) attributed to Hyginus, the librarian of Augustus and author of the *Fabulae*, a compendium of classical myths. The date and attribution of the *Poetic Astronomy* have both been contested; however, a recent editor of the *Poetic Astronomy* argues convincingly that similarities in content between it and the *Fabulae*, along with the absence of astrological allusions in the *Poetic Astronomy*, point both to a common authorship and to a date of composition before astrology became fashionable in Rome, i.e., a few years B.C.E. If that date is accurate, then Hyginus's work may have antedated *The Constellations*, and his repeated citing of Eratosthenes as source in the *Poetic Astronomy* may well be a reference to the original *Catasterismi* of Eratosthenes.

The earliest Greek references to constellations are found in Homer, who describes as follows the intricate decorative scenes depicted on the shield that the god Hephaestus forged for Achilles:

> He made the earth upon it, and the sky, and the sea's water,
> and the tirelesss sun, and the moon waxing into her fullness,
> and on it all the constellations that festoon the heavens,
> the Pleiades and the Hyades and the strength of Orion
> and the Bear, whom men give also the name of the Wagon,
> who turns about in a fixed place and looks at Orion
> and she alone is never plunged in the wash of the Ocean.
> On it he wrought in all their beauty two cities of mortal
> men. And there were marriages in one, and festivals . . .

But around the other city were lying two forces of armed men
shining in their war gear. For one side counsel was divided
whether to storm and sack, or share between both sides the property
and all the possessions the lovely citadel held hard within it.

(Iliad 18.483–92, 509–12. tr. R. Lattimore)

The depiction of earth, sea, and sky on the shield of Achilles is
comprehensive in scope. The human activity of the two cities is
described in such minute detail that it is tempting to take Homer at his
word when he represents "all the constellations that festoon the
heavens" as being four in number, namely, the Pleiades, Hyades, Ursa
Major and Orion. A passage in the Odyssey mentions the Pleiades,
Ursa Major, and Orion, and also refers to "late-setting Bootes";
however, given the context—Odysseus is sailing by the stars, as
instructed by Calypso—it is not clear whether "Bootes" refers to the
constellation Bootes or to its brightest star, Arcturus.

Glorious Odysseus, happy with the wind, spread sails
and taking his seat artfully with the steering oar he held her
on her course, nor did sleep ever descend on his eyelids
as he kept his eye on the Pleiades and late-setting Bootes,
and the Bear, to whom men give also the name of the Wagon,
who turns about in a fixed place and looks at Orion,
and she alone is never plunged in the wash of the Ocean.

(Odyssey 5.269–75. tr. R. Lattimore)

There is no explicit reference to constellation myths in Homer;
however, there are two oblique references, both with reference to the
Bear (Ursa Major), which is said "to keep a watchful eye" on Orion,
who as a hunter is presumably on the lookout for prey.[14] Homer also
refers to the fact that the Bear, uniquely, does not set—i.e., is always
visible above the horizon—implying that there is a reason for this
unique phenomenon.[15]

Hesiod refers to the same constellations as Homer, citing their rising or setting as the appropriate signal to undertake certain tasks such as harvesting or ploughing, pruning, harvesting grapes and making wine, or sailing.

> Start reaping when the Pleiades rise, daughters of Atlas,
> and begin to plow when they set.
>
> (Hesiod, *Works and Days*, 383–84, tr. A. N. Athanassakis)

> When—Zeus willing—counting from the winter solstice
> sixty days have passed, then the star Arcturus
> leaves the sacred stream of Okeanos
> and first rises brilliant at eventide,
> then . . . it is best to prune your vines . . .
>
> (*Works and Days*, 564–70)

> When Orion and the dog star rise to the middle of the sky
> and rosy-fingered dawn looks upon Arcturus,
> then, Perses, gather your grapes and bring them home . . .
> When the Pleiades, the Hyades, and mighty Orion set,
> remember the time has come to plow again . . .
> When the Pleiades flee mighty Orion
> and plunge into the misty deep
> and all the gusty winds are raging,
> then do not keep your ship on the wine-dark sea.
>
> (*Works and Days*, 609–22)

In addition to the four constellations of the Pleiades, Hyades, Ursa Major, and Orion, Homer and Hesiod both refer to the individual stars Arcturus and Sirius. Thus it would seem safe to say that the Greeks of the eighth and seventh centuries B.C.E. distiguished at least four constellations—or five, if Homer's Bootes is counted—and several individual stars. There was a tradition, referred to by Hyginus, that the constellation Ursa Minor was introduced by Thales of Miletus in the sixth century B.C.E., and it is clear that Coma Berenices was added in the third century B.C.E., and Antinous in the second century C.E. Concerning the remaining constellations, it is impossible to say when they became part of the Greek sky.

The Constellations cites Hesiod as the authority for several of the myths recounted, but modern scholars have been for the most part unable to detect any traces of constellation myths in the extant works of Hesiod, or to determine whether any such myths were included in the lost work entitled *Astronomia*, which is attributed to Hesiod. Similary, it is difficult to know the precise content of lost works by early authors such as Epimenides and Pherecydes, who are cited both in *The Constellations* and the *Poetic Astronomy* as sources. But even in those instances when the reference is to an existing work, such as a play of Euripides or Aeschylus, it does not appear that the references in *The Constellations* point to anything beyond the myth itself; i.e., the ending of the myth as recounted in the source cited by the *The Constellations* does not entail the changing of the personages in the myth into constellations. With one exception (the myth related about the constellation Piscis Austrinus), the myths recounted in the *The Constellations* are familiar Greek myths that are well-attested in classical literature. What is unclear is when and how these myths became associated with a particular constellation. Either the linking of myth and constellation was a literary construct invented by Eratosthenes in his *Catasterismi*, or, more likely, it evolved slowly in popular imagination during the centuries between Homer and Hesiod and the Hellenistic age.

On the other hand, there are certain myths, attested only in literature similar to *The Constellations*, which most scholars believe to have originated from the relative position and movement of two or more constellations in the sky. Such myths, called astral or astronomical myths, may include the death of Orion from the sting of the Scorpion; the "heavenly hunt scene" consisting of Orion, his dogs (Canis Major, Canis Minor), and his game (Lepus); the pursuit of the Pleiades by Orion; and, according to one scholar, the Perseus-Andromeda story.

Anyone who has confronted the starry sky on a moonless night, away from the lights of civilization, can begin to imagine how the nightly appearance of the stars and their regular risings and settings through the course of the year might invite speculation. Certain groupings of stars, such as those constituting Orion or Ursa Major, stand out from the rest; other less prominent groupings such as the Pleiades or the Hyades are, nevertheless, easily distinguished and are useful as seasonal markers. But there are many more stars in the sky than those comprising the four constellations just mentioned, and many of those stars can be grouped together. And if Orion, the Pleiades, and the Hyades are familiar as mythological figures, why cannot other mythological figures also be represented among those other groups of stars? Furthermore, if one constellation rises when another sets, might there not be a connection between the two? Some such speculation might have resulted in populating the Greek sky with the plethora of mythological birds, beasts, heroes, and inanimate objects that some Greek deity or deities saw fit to honor by placing them among the stars. Some of the constellations form groups comprising a common mythological theme: e.g., the five constellations of the Perseus-Andromeda group, and the five constellations mythologically connected with Orion: Scorpio, Canis Major, Canis Minor, Lepus, and the Pleiades.

The constellation figures of the Greeks included heroes of mythology (Heracles, Perseus, Orion, Castor, Polydeuces and Asclepius), nymphs (Hyades, Pleiades), animals and birds (two bears, three serpents, a scorpion, two dogs, a crab, lion, goat, bull, horse and ram, three fish, a swan, an eagle, a crow, hare, dolphin and sea-monster), and inanimate objects (a crown, triangle, lyre, arrow, altar, crater, river, ship, and lock of hair). Most of these figures were known to the Babylonians and some were known to the Egyptians (only Coma Berenices and Antinous can be dated with certainty to the Hellenistic and Roman periods, respectively), but there is little correspondence between Babylonian, Egyptian, and Greek constellations, either in their location or in their delimitation.

The earliest surviving artistic representation of the constellations known to the Greeks and Romans is the Atlas Farnese, a marble globe depicting the five heavenly circles and most of the constellation figures, but not the individual stars comprising each constellation. The Atlas Farnese is variously dated between the second century B.C.E. and about 150 C.E. and appears to be based on Hipparchus.[16] The only other pre-modern representations of the constellation figures are twenty-nine illustrated manuscripts dating from the ninth century and later, the chief of which are Vaticanus Graecus 1291 and Vaticanus Graecus 1087. According to ancient tradition, Thales of Miletus, Anaximander, and Eudoxus constructed *sphaerae*, i.e., graphic representations of the heavens, but these *sphaerae* have been lost.

Andromeda

The Constellations 17

Andromeda was placed among the stars by Athena to commemorate the feats of Perseus. Her arms are outstretched as when she lay exposed to the sea monster. After being rescued by Perseus, Andromeda, being nobly minded, chose not to return to her parents, but of her own accord followed Perseus to Argos. Euripides tells the story in detail in the play he wrote about her.

Andromeda has one bright star on the head [α]; one star on each shoulder [π, ε]; one on the right elbow [σ?]; one bright star at the edge of the hand [ι?]; one on the left elbow [η]; one on the arm [ρ?]; one on the hand [ο]; three on the belt [?, ?, ?]; four above the belt [β, μ, ν, ?]; one bright star on each knee [υ?, φ]; two on the right foot [φ Per, 51]; and one on the left foot [γ]. The total is twenty.

Poetic Astronomy 2.11

This figure was reportedly placed among the stars by Minerva because of Perseus's courage, for he saved Andromeda from danger when she was exposed to the sea monster. Nor did Perseus receive lesser consideration from Andromeda; for neither her father Cepheus nor her mother Cassiopeia could prevail upon her not to leave her country and follow Perseus. Euripides has written fully about her in his play of the same name.

3.10

As we noted earlier [2.11], she has one bright star on the head; one on each shoulder; one on the right elbow; one in her hand; one on the left elbow; one on the arm; another in the other hand; three on her belt; four over the belt; one on each knee; two on her feet. Thus the total is twenty stars in all.

Commentary

Andromeda was the daughter of Cepheus and Cassiopeia and the innocent victim of her mother's arrogance. According to most accounts, after her rescue from the sea monster, Andromeda became the wife of her deliverer, Perseus. Perseus and Andromeda stayed with King Cepheus for a time, then went to Argos, leaving their first-born son, Perses, to succeed Cepheus, who had no sons. At Argos or Tiryns, Andromeda bore five more sons and a daughter.[1]

The folk-motif of the maiden in distress who is rescued and marries her rescuer is widespread, as is that of the monster-slaying hero.[2]

The origin of this constellation appears to be Phoenician. The name Andromeda itself is a part-translation, part-transliteration of the Phoenician name Adamath.[3]

The number of stars comprising Andromeda is twenty according to Ps-Eratosthenes, Hyginus, and Hipparchus, twenty-three according to Ptolemy.

Aquarius

The Constellations 26

This figure is thought to have been named the Water-Pourer [Aquarius] because of the action he represents. The Water-Pourer stands holding a wine-jar, from which he is pouring a stream of liquid. Some find in this image sufficient proof that the figure represented is Ganymede, and they call Homer to witness, because the poet says that Ganymede, adjudged worthy by the gods, was carried away on account of his beauty to be cup-bearer to Zeus. Homer says, too, that Ganymede was granted immortality, which was yet unknown to men. The liquid being poured from the jar is said to resemble nectar, the drink of the gods, and this resemblance is interpreted as proof that the liquid is indeed the aforementioned drink of the gods.

Aquarius has two faint stars on the head [25, ?]; one star on each shoulder [α, β], both large stars; one on each elbow [γ, ν?]; one bright star at the edge of the right hand [π?]; one on each breast [ξ, ?]; one under the breast on either side [ε, ?]; one on the left hip [ι?]; one on each knee [τ, 66]; one on the right leg [δ]; one on each foot [68, ?]. The total is seventeen. The stream of water consists of thirty-one stars, of which two stars are bright [κ, α PsA].

Poetic Astronomy 2.29

Many say this is Ganymede who, on account of his beauty, was snatched away from his parents by Jupiter to be the cup-bearer of the gods. Thus he is represented as pouring water into some object.

Hegesianax, however, says the figure is Deucalion because, during his reign, such a quantity of water fell from heaven that a great flood reportedly occurred. Eubulus says the figure is Cecrops, citing the antiquity of his lineage and pointing out that Cecrops reigned before wine was invented, and that before wine was known to man, water was used in sacrificing to the gods.

3.28

The figure has two faint stars on the head; one bright star on each shoulder; one large star on the left elbow; one on the right; one on the front hand; one faint star on each breast; two under the breast; one on the inner thigh; one on each knee; one on the right leg; one on each foot. There are seventeen in all. The flow of water, including the water-jar, consists of thirty-one stars of which the first and northernmost are bright.

Commentary

This constellation is generally associated with water, not only in Greek literature but also in ancient Egyptian and Babylonian records. The three figures identified with Aquarius by classical authors are the cup-bearer Ganymede; Deucalion, the hero of the Greek flood story; and Cecrops, an early king of Athens. The constellation may have been associated with water because of its appearance during the rainy season. Aquarius was the eleventh zodiacal sign in Euphratean records and coincided with the winter solstice at the time when the vernal equinox was in the constellation Taurus.[1]

The story of Ganymede attributed by *The Constellations* to Homer is an accurate paraphrase of *Iliad* 20.231–35:

and to Tros in turn there were born three sons unfaulted,
Ilos and Assarakos and godlike Ganymedes
who was the loveliest born of the race of mortals, and therefore
the gods caught him away to themselves, to be Zeus' wine-pourer
for the sake of his beauty, so he might be among the immortals.

(tr. Lattimore)

In addition to the above traditional account of Ganymede's abduction, two other traditions speak of Ganymede's abduction by Tantalus or Minos. Some sources mention a gift—consisting either of horses or a golden vine fashioned by Hephaestus—presented to Tros by Zeus as compensation for the loss of Ganymede. The changing of Ganymede into a constellation is recounted only in late sources.[2]

The constellation Aquarius is probably of Euphratean origin and may be represented on Babylonian boundary-stones as a man or boy pouring water from an urn. In Egypt, this sign was called "Water," and its hieroglyph (≋) is still used as the symbol for Aquarius. In classical times, the figure was usually depicted in a standing position, holding a jar from which a stream of liquid poured forth.[3]

According to Ps-Eratosthenes and Hyginus, the number of stars comprising Aquarius is forty-eight, of which seventeen comprise the figure of the man and thirty-one the liquid; Hipparchus counts eighteen stars, and Ptolemy forty-two.

Aquila

The Constellations 30

This is the eagle which brought Ganymede to heaven to be the cup-bearer of Zeus. The eagle is among the stars because earlier, when the gods were casting lots for the various birds, the eagle fell to Zeus. It is the only bird which flies toward the sun, not bowing to the sun's rays, and it holds first place among the birds. This constellation represents the eagle with wings outspread as if in downward flight. Aglaosthenes says in his *Naxica* that Zeus, after his birth on Crete, was sought out [by his father] and twice carried away. He was subsequently removed from Crete and brought to Naxos, where he was raised. When he came of age, he gained the overlordship of the gods. As he was leaving Naxos to go against the Titans, an eagle appeared at his side. Zeus accepted the omen and adopted the eagle as his own bird. For this reason, the eagle was deemed worthy of honor in the heavens.

The Eagle is comprised of four stars [α, β, ζ, τ] of which the central star [α] is bright.

Poetic Astronomy 2.16

This is the eagle said to have snatched up Ganymede and delivered him to his lover, Jupiter. Jupiter was said to have singled out this bird from all the race of birds. According to tradition, this is the only bird that tries to fly against the rays of the rising sun; it appears to be flying above Aquarius, for many identify that constellation with Ganymede.

Some say that there was a certain Merops, who reigned over the island of the Coans, and that he named the island Cos after his daughter, and its inhabitants Meropians, after himself. Merops had a wife, Ethemea, who was descended from the race of nymphs. When she ceased to worship Diana, Diana shot her with arrows, but she was carried away, still alive, by Proserpina to the Underworld. Merops, moved by longing for his wife, wished to kill himself, but Juno pitied him and placed him among the stars, transforming his body into an eagle—for if she had placed him there in the shape of a man, he would have retained his human memory and continued to long for his wife.

However, Aglaosthenes, who wrote the *Naxica*, says that Jupiter was spirited away from Crete and carried to Naxos where he was raised. Later, when he attained manhood and wished to destroy the Titans in war, an eagle appeared to him while he was sacrificing; he accepted this omen and placed the eagle among the stars. Some, however, say that Mercury—others claim it was Anaplades—struck by the beauty of Venus, fell in love with her and, when she would give him no chance, became dispirited, as if having been disgraced. Jupiter pitied him and, when Venus was bathing in the river Achelous, sent an eagle that carried away her sandal to Amythaonia in Egypt and handed it over to Mercury. Seeking her sandal, Venus came to the one who desired her, and he, on obtaining what he desired, placed the eagle in the firmament in exchange for the service rendered.

The figure has one star on the head [τ]; one on each wing [α, β]; one on the tail [ζ].

Commentary

In both the Greek and Latin traditions, primary identification of this constellation is with the eagle of Zeus. Indeed, Ps-Eratosthenes devotes most of his narrative to how the eagle came to be associated with Zeus.

The connection of Aquila with the story of Ganymede may have been a later embellishment, influenced by the proximity of Aquila to the constellation Aquarius, which was identified with Ganymede. The earliest references to the story of Ganymede make no mention of an eagle. In Homer, for example, Ganymede is said to have been snatched away by the gods because of his beauty to be the cup-bearer of Zeus. The introduction of the eagle into the story dates from the fourth century B.C.E.[1]

The association of the king of birds with the father of the gods is not surprising. The primacy of place among birds ascribed to the eagle by the Greeks and Romans parallels the status accorded to the lion among four-footed animals, and, in both instances, is cited by Ps-Eratosthenes to explain the origin of the relevant constellation.[2]

Hyginus's account agrees with that of Ps-Eratosthenes, but offers, in addition, two alternative explanations for the constellation: in one, the eagle serves as the messenger of Zeus; the other is remarkable in that it omits any connection between the eagle and Zeus, attributing the origin of the constellation to Juno.[3]

During the reign of the emperor Hadrian, an attempt was made to form a constellation out of six or seven stars in the lower part of the constellation Aquila (δ, η, θ, ι, κ, λ, ν Aql). The new constellation was to be called Antinous, in honor of the emperor's favorite who had recently drowned in the Nile River. Ptolemy, who compiled his star-catalogue twenty to thirty years after the reign of Hadrian, lists six stars of Antinous among the "unformed" stars of Aquila. The constellation Antinous persisted into modern times, appearing in star tables until the late eighteenth century, but its stars are now included among those of Aquila.[4]

The constellation of the Eagle may be of Euphratean origin. A Mesopotamian cylinder seal, believed to illustrate an episode from the Gilgamesh Epic, depicts a figure ascending to heaven on the back of an eagle.[5]

The Eagle consists of four stars according to Ps-Eratosthenes and Hyginus, five according to Hipparchus, and nine according to Ptolemy.

Ara

The Constellations 39

This figure is the altar upon which the gods first swore their allegiance, at the time when Zeus waged his campaign against Cronus. The altar was constructed by the Cyclopes and had a cover over the flame so that the strength of the lightning bolt might not be visible [to Cronus]. When Zeus and the gods emerged as victors, they introduced altars among men so that those swearing allegiance to one another might sacrifice upon them. In athletic contests and in [. . .], those wishing to pledge a most sacred trust touch the altar with their right hand, considering this to be a sign of good will. Likewise, seers sacrifice upon an altar when they wish to see most clearly.

The Altar [Ara] has two stars on the altar-pan [α, β] and two on the base [σ, θ]. The total is four.

Poetic Astronomy 2.39

On this altar, the gods were believed to have first made offering and taken oaths, when they were about to fight against the Titans. The altar was made by the Cyclopes. Men were said to have adopted this custom, so that when they have in mind a plan of action, they offer sacrifice before undertaking it.

3.38

There are two stars at the top of the incense-pan and two at the bottom, four altogether.

Commentary

The principal textual sources for the battle of the Olympian gods against the Titans (Titanomachia) are Hesiod, Apollodorus, and Claudian. Few and often incomplete as they are, these sources provide no allusion to the pact of the gods mentioned by Ps-Eratosthenes and Hyginus. Thus, it appears that the author of *The Constellations* either drew his information from a source no longer extant, or, faced with an Eastern constellation figure for which no appropriate Greek myth could be found, resorted to his own imaginative powers.[1]

This constellation, which is always identified as an altar or a censer in classical literature, is explained in three different ways: as the altar on which the gods swore their pact, as the altar upon which Centaurus is sacrificing, or as the altar used at the marriage of Peleus and Thetis. The references to the constellation in classical literature as both an altar (*thyterion*) and a censer (*thymiaterion, turibulum*) suggest that the prototype of the constellation figure was an altar-censer of the type which has been found in the Near East. Excavations in Greece have produced altars and censers with firepans (i.e., metal layers of the same shape as the top of the altar which received the sacrificial fire and protected the altar from the flame), but not with covers.[2]

The origin of the Greek constellation Ara is probably to be found in the Euphratean altar-constellation. However, the position of the Greek constellation did not correspond to that of its Euphratean counterpart, since the latter was located in the Claws of the Scorpion (Libra).[3]

The Altar is comprised of four stars according to Ps-Eratosthenes, Hyginus, and Hipparchus, seven according to Ptolemy.

Argo

The Constellations 35

The Argo was placed among the stars at the wish of Athena, as a clear example to later generations of men. It was the first ship to be built and to cross the as yet untraversed sea. It was also endowed with speech. Its image was not placed among the stars in its entirety; the helm is visible up to the mast, along with the steering-oars, so that those at sea who look upon it may be encouraged to labor, and so that its glory might be ageless, as it is among the gods.

The Argo has four stars on the stern [11, ρ, π, ?]; five on one steering-oar [η Col, ν Pup, ?, ?, ?] and four on the other [α, τ Pup, ?, ?]; three on the edge of the stern-mast [α Pyx, β Pyx, γ Pyx]; five on the deck [f Pup, BSC2961, c Pup, b Pub, ζ Pup]; six close together under the keel [χ, o Vel, δ Vel, f, K Vel, N Vel]. The total is twenty-seven.

Poetic Astronomy 2.37

Some say this ship was called *Argo* ["swift"] by the Greeks because of its speed, others because Argus was its builder. Many say that this was the first ship to sail the seas and that it came to be represented among the stars principally for the following reason. Pindar says this ship was built in the town of Magnesia called Demetrias; Callimachus

says it was built in that same region, in the place called Pagasae—
pagasai ["entrance"] in Greek, because the Argo was first fitted
together there—near the temple of Apollo Actius, past which the
Argonauts sailed as they were setting out. Homer locates Pagasae in
Thessaly. Aeschylus and others, however, say that a speaking plank
was added to the ship by Minerva. The ship's form is not entirely
visible among the stars: it is represented from stern to mast, signaling
that men should not be greatly afraid when their ships are torn
asunder.

3.36

The Argo has four stars on the stern; five stars on the first steering-
oar and four on the other; six around the keel; five above the deck;
three on the mast. Thus the total number of stars is twenty-seven.

Commentary

The legendary voyage of the Argo in quest of the Golden Fleece was
a favorite subject of both Greek and Latin authors. Whether the Argo
was the first ship to sail the Greek sea is uncertain—some ancient
sources contended that Minos's navy and the ship of Danaus antedated
the Argo. What appears certain is that the Argo was the first Greek
ship to be launched and that the quest for the Golden Fleece was the
first Greek naval expedition. The inspiration for the building of the
Argo was provided by Athena in her capacity as goddess of knowledge
and skill. In most accounts, the ship was built by Argus, the son of
Phrixus, with wood from Mount Pelion, and was manned by Jason and
a crew of fifty Greek heroes. The Argo carried Jason and his Argonauts
("sailors of the Argo") from Iolcus to Colchis, where they succeeded
in capturing the Golden Fleece, then brought them back to Greece.
According to the usual account, the Argonauts on their return dedi-
cated the Argo to Poseidon at the Isthmus of Corinth.[1]

Like most legends, the Argonautic expedition slowly acquired an
aura of folktale so that the Argo eventually came to be represented in

classical literature as a magic ship, having in her prow a piece of wood from the oak groves of the oracle of Zeus at Dodona. This piece of wood possessed both the power of speech and the ability to see into the future, and was instrumental in ensuring the safety of ship and crew at various points along their journey. The all-star crew of Greek heroes and the great swiftness of the Argo also may belong to the realm of folktale.[2]

Most of the references to this constellation in classical literature describe the figure represented as the rear half of a ship. Several reasons are given for the incompleteness of the figure. *The Constellations* contends that this part of the ship would inspire those sailors who looked upon it to work more zealously. Hyginus states that the sight of half a ship would encourage seafarers not to fear shipwrecks. A third reason given is that the missing part of the constellation-figure represented that part of the Argo which was crushed by the Symplegades. None of the above explanations appears authoritative, nor is any one of them attested by more than one source. The first two explanations appear to be no more than imaginative attempts on the part of their authors to moralize. The third explanation ignores the details of the Argonautic legend: the Argo was in fact damaged while passing through the Symplegades ("Clashing Rocks"), but the damage was to its stern, not its prow, and was very minor. There is no basis in literary sources for the contention that half the ship was crushed. Two possible explanations have been advanced by modern scholars for the incompleteness of this constellation-figure: 1) Argo may belong to that group of incomplete constellation-figures including Taurus and Pegasus, 2) the Greeks somehow associated the constellation Argo with a seventh-century Phoenician warship, which to them resembled half a ship. The first explanation still fails to explain why the Greeks, given the "complete" form of the Euphratean constellation-figures, would represent their own figures as incomplete. The second explanation seems more plausible. If the constellation Argo came to the Greeks through the Phoenicians, as several constellations did and the adoption of this constellation by the Greeks occurred in the seventh or sixth century B.C.E., as was the case with the majority of the constellations, then it is possible that the constellation figure of a ship would be represented as a contemporary Phoenician ship. The Phoenician war galleys of the seventh century B.C.E. were of such a shape as to resemble the rear half of a ship to the Greeks, who had never built ships with the prow ending in a vertical line.[3]

This constellation is always identified with a ship, usually the Argo, but also the ship of Danaus or of Osiris. The origin of this constellation is probably to be sought in the Euphratean "Ship of the Canal of Heaven."[4]

The number of stars in Argo is twenty-seven according to Ps-Eratosthenes and Hyginus; twenty-six according to Hipparchus, and forty-five according to Ptolemy. The brightest star (α Car) was called Canopus or Perigeios by the Greeks, but may not have been visible from mainland Greece in antiquity.[5]

Aries

The Constellations 19

This is the ram that carried Phrixus and Helle. The ram was immortal and was given to them by their mother Nephele ["cloud"]. According to Hesiod and Pherecydes, the ram had a golden fleece. As the ram was carrying them across the strait called Hellespont after Helle, he threw Helle off, and also lost his horn—Helle was rescued by Poseidon who fathered a son, Paeon, by her—but carried Phrixus safely to the realm of Aeetes on the Euxine Sea. There the ram shed the golden fleece and gave it to Phrixus as a remembrance, then went away to the stars, whence he appears very faint.

The Ram has one star on the head [α]; three on the nostrils [η, θ, ?]; two on the neck [ι, ?]; one bright star on the edge of the front leg [σ?]; four stars on the back [ε, ρ?, 41?, ?]; one on the tail [δ]; three under the belly [ρ?, σ?, μ?]; one on the haunch [ν]; one on the edge of the back leg [μ Cet]. The total is seventeen.

Poetic Astronomy 2.20

This constellation is said to be the ram that carried Phrixus and Helle across the Hellespont. Hesiod and Pherecydes say he had a golden fleece—we will speak about this elsewhere. Now, Helle fell into the Hellespont and, ravished by Neptune, gave birth to Paeon, according to some, or to Edonus, according to others. After arriving safely in the kingdom of Aeetes, Phrixus sacrificed the ram to Jupiter and hung the fleece in the temple. The image of the ram was placed among the stars by Nubes ["cloud"] to preside over the time of year when the grain is sowed because formerly Ino sowed parched grain at that time—which was largely responsible for the flight [of Phrixus and Helle]. Eratosthenes says that the ram itself removed its golden fleece and gave it to Phrixus as a remembrance, then flew away to the stars, for which reason, as we mentioned earlier, it shines faintly.

Some say that Phrixus was born in the town of Orchomenus, which lies in Boeotia, others at Salonus in the region of Thessaly. Others say that Cretheus and Athamas, among many others, were sons of Aeolus; others that Salmoneus was the son of Athamas and grandson of Aeolus. Cretheus had as his wife Demodice, whom others call Biadice. She fell in love with Phrixus, the son of Athamas, on account of his beauty, but could not prevail upon him; consequently, forced by the circum-stances, she accused him before Cretheus of attempting to rape her, and said other like things, as women will. Cretheus, as became a loving husband and a king, grew angry and persuaded Athamas to put Phrixus to death. However, Nubes intervened and snatching away Phrixus and his sister Helle, placed them on a ram, and ordered them to flee across the Hellespont as far away as possible. Helle fell off and drowned, and the sea was named Hellespont after her, but Phrixus came to Colchis and, as we mentioned earlier, hung the fleece of the sacrificed ram in the temple. He himself was brought back to Athamas by Mercury, who satisfied Athamas that Phrixus had fled trusting in his own innocence.

Hermippus, however, says that when Liber was campaigning in Africa, he came with his army to a place called Ammodes ["sandy"] because of the large amount of sand there. He was in great danger because he saw that he had to advance, and that water was in very short supply. Because of this the army was reaching exhaustion. As they were considering what to do, a ram, wandering alone, chanced to come before the soldiers, fleeing when it saw them. The soldiers, however,

had seen it, and, although they were advancing with difficulty, op-
pressed by dust and heat, nevertheless followed the ram, as if seeking
a prize from the flames, to that place named after the temple to Jupiter
Ammon, which was later built there. When they came to that place,
they could nowhere find the ram which they had followed, but—what
was more to be desired—they found a great quantity of water. They
refreshed themselves and straightway brought the news to Liber. He,
rejoicing, led his army to that region and built a temple to Jupiter
Ammon in which the statue of the god is depicted with ram's horns.
The ram he represented among the stars such that when the sun is in
its sign, all things that grow are renewed; this comes to pass in the
spring because of the fact that Liber's army was renewed by the ram's
flight. Furthermore, Liber wished that it should be first among the
twelve signs because it was the best leader of his army.

Concerning the image of Ammon, Leon, who wrote about Egyptian
matters, says that when Liber was ruling over Egypt and the other
lands and introduced all arts to mankind, a certain Ammon came out
of Africa and brought a great flock of sheep to Liber, in order to gain
his favor and to be recognized as having invented something. And so,
Liber reportedly gave him as a reward the land opposite Thebes in
Egypt, and those who make statues of Ammon show him with horned
head, so that men might remember that he was the first to make use
of sheep. Those who wish to assign this gift to Liber, as not sought
from Ammon but offered of his own accord by Ammon to Liber, make
horned images of Liber and say he placed the ram among the stars as
a memorial.

3.19

The Ram has one star on the head; three on the horns [β, γ, ?]; two
on the neck; one on the edge of the front foot; four on the back; one
on the tail; three under the belly; one on the flank; and one on the back
foot. The total number of stars is seventeen.

Commentary

The figure in this constellation is usually connected with several different rams in the Greek and Roman tradition. The most common identification, related by both Ps-Eratosthenes and Hyginus, is with the golden-fleeced ram of Phrixus and Helle. Hyginus relates two additional versions, according to which the ram was changed into a constellation by Liber during his Egyptian campaign, either because it led the army to water, or because it commemorates the invention of sheep-breeding. Aries is also identified with the golden lamb of Thyestes and Atreus. A late source identifies the constellation as Athena seated upon or beside a ram. The placing of Athena in this constellation may reflect an attempt to add the goddess to the nearby Perseus-Andromeda group, with which she is mythologically connected.[1]

Phrixus and Helle were the children of Athamas, king of Boeotia, by the goddess Nephele (Latin "Nubes"). In the usual account, Athamas abandoned Nephele for Ino, who devised a plot against Phrixus and Helle. She caused the wheat crop to fail by persuading the women to parch the wheat seeds before sowing them, then bribed the messenger sent to consult the oracle at Delphi to announce on his return that disaster could be averted only through the sacrifice of Phrixus. Athamas was about to sacrifice Phrixus, when either Nephele or Zeus himself sent a golden ram to carry the children to safety.[2]

After the fall of Helle, the ram brought Phrixus to Colchis. There, Phrixus either sacrificed the ram and presented the golden fleece to King Aeetes, or the ram shed the fleece of its own accord and flew away to the heavens to become a constellation. Phrixus was well received by Aeetes, who gave him his daughter Chalciope in marriage.[3]

Aeetes was the son of Helius, and the brother of Circe and Pasiphae. He was also the father of Medea. The entire family possessed powers of sorcery and produced the few sorceresses of Greek mythology.[4]

The story of Phrixus probably reflects an ancient custom of sacrificing a king's son in time of famine. The story was current in the area around Mount Laphystion in Boeotia and also in Thessaly at the city of Alus (Halus), which was said to have been founded by Athamas. In both places, human sacrifice was offered to Zeus Laphystius, the sacrificial victim being the oldest son of the priestly family, which in both cases was descended from Athamas. The practice of offering human sacrifice survived at Alus until at least the fourth century B.C.E. At a later date, the human victim was replaced by a ram which was sacrificed by a member of the priestly family. The ram was the most common sacrificial animal in the cults of Zeus.[5]

The role of Helle in the story is insignificant. She serves solely as the eponym of the Hellespont. In the usual account, Helle fell off the ram as it flew over the strait between Europe and Asia. The tradition that she was thrown off by the ram, who also lost his horn in the process, occurs only in *The Constellations*.[6]

Hyginus is the only ancient author to connect Liber with this constellation. The oracle of Zeus-Ammon in the Libyan desert was known to the Greeks at the time of Herodotus. Liber's campaign in Egypt and his arrival at the site of the temple of Zeus-Ammon after a long and thirsty march is reminiscent of an expedition undertaken by Alexander the Great to the same site. After consulting the oracle, Alexander proclaimed himself the son of Zeus-Ammon. Subsequently, Alexander was portrayed on numerous coins wearing ram's horns.[7]

Paeon, the son of Poseidon and Helle, is to be identified with the eponymous ancestor of the Paeonians, although some ancient sources give a different genealogy to the Paeonians.[8]

Aries consists of seventeen stars according to Ps-Eratosthenes, Hyginus, and Hipparchus, thirteen according to Ptolemy, who lists no star brighter than the third magnitude.

Auriga

The Constellations 13

This is Erichthonius, the son of Hephaestus and Gaea, who was the first man to harness horses to a chariot. Zeus saw him and marvelled that he imitated Helius in building a chariot and harnessing four white horses to it, and thus placed him among the stars. Erichthonius was the first to lead a procession to the Acropolis in honor of Athena. In addition, he made her rites famous by exalting her statue. Euripides tells the following story of Erichthonius's birth. Hephaestus became enamored of Athena and wished to have intercourse with her, but she resisted, preferring her chastity, and hid at a place in Attica which was called Hephaestium, after him. Hephaestus, thinking she was overcome, approached her, but Athena wounded him with her spear and his passion subsided, his seed falling upon the earth. From this seed arose a boy called Erichthonius ["Earth-born"], from the manner of his birth. When Erichthonius reached manhood, he invented the chariot and won acclaim as an athlete. He participated studiously in the Panathenaean games, driving a chariot with a rider beside him who carried a small spear and wore a helmet. From this rider [Gk. *parabates*], the *apobates* received his name.

Within this constellation are the Goat [Capella] and the Kids

49

[Haedi]. According to Musaeus, when Zeus was born, Rhea gave him over to the keeping of Themis, and Themis in turn gave the infant to Amalthea. Amalthea put him with her goat, which nursed Zeus. This goat was the daughter of Helius and was so ugly that the gods of Cronus's time, in disgust at her appearance, ordered Gaea to hide her in one of the caves on Crete. Having hidden the goat, Gaea appointed Amalthea to watch over her, and Amalthea nursed Zeus with the milk of this goat. When Zeus came of age and was preparing to attack the Titans, he possessed no weapons. An oracle told him to use the skin of the goat as a shield since it was impenetrable and terrible to look at, having a Gorgon's head in the middle of the back. Zeus did as he was bidden and by this stratagem appeared twice his actual size. It is said he changed the goat into a star, after covering her bones with another skin and granting her life and immortality [. . .].

Some say this constellation is Myrtilus, the son of Hermes and charioteer of Oenomaus.

The Charioteer has one star on the head [δ]; one on each shoulder [α, β], of which the left one, which is called the Goat [Capella], is bright; one star on either elbow [ε, ν]; one on the right hand [θ]; two on the left hand [η, ζ], which are called the Kids [Haedi]. The total is eight.

Poetic Astronomy 2.13

According to Eratosthenes, this figure is Erichthonius; we call him Auriga ["charioteer"] in Latin. When Jupiter saw that Erichthonius was the first to yoke horses to a four-horse chariot [*quadriga* in Latin], he marvelled at the ingenuity of the man in approaching the inventions of the Sun God, as the latter was first among the gods to make use of a four-horse chariot. Erichthonius was the first to utilize the quadriga, as we mentioned earlier, to institute sacrifices to Minerva, and to dedicate a temple on the acropolis of Athens.

Concerning Erichthonius's birth, Euripides recounts that Vulcan, struck by the physical beauty of Minerva, sought her in marriage but did not obtain her consent; Minerva went to hide in the place which was subsequently known as Hephaestium because of these events. Vulcan, it is said, gave chase with mounting passion, and when, greatly aroused, he tried to embrace her, he was repulsed and his seed fell upon the earth. Minerva, overcome by shame, fled from the place. However, the serpent Erichthonius was born of that union, and has that name because he was born from the earth and from their dissension. Minerva placed the infant in a little chest and gave him over to the care of the daughters of Erechtheus as a secret charge, ordering them not to open the chest. But, human nature being greedy, and very often desirous of what is most prohibited, the maidens opened the chest and saw the serpent. Because of this, Minerva drove them mad and they hurled themselves from the Acropolis. The serpent took refuge in the shield of Minerva and was raised by her.

Others say that Erichthonius had the lower members of a serpent, and that as a youth he competed in the games of Minerva Panathenaea with the four-horse chariot; because of this he was reportedly placed among the stars. Others who have written about the constellations say this is Orsilochus, an Argive by birth, who first introduced the quadriga and received a place among the stars on account of his invention.

Others, however, say this is Myrtilus, the son of Mercury by Clytia, and charioteer of Oenomaus. After his memorable death, his father placed his body in the heavens. On his left shoulder lies the Goat and in his left hand the Kids are visible. Concerning these, some say the following. There was a certain Olenus, a son of Vulcan and father of two nymphs, Aega and Helice, who were nurses of Jupiter. Others say that certain cities were named after them including Olenus in Elis, Helice in the Peloponnesus, and Aega in Haemonia. Homer speaks of these in the second book of the *Iliad*.

Parmeniscus, however, said that there was a certain Melisseus, king of Crete, and that Jupiter was entrusted to his daughters to raise. When these lacked milk, they brought him to a goat named Amalthea, who reportedly raised him. This goat used to give birth to twin kids and had given birth when Jupiter was brought to her to suckle. And so, because of the services of the mother, the kids also were placed among the stars. Cleostratus of Tenedos was the first to identify the Kids among the stars.

Musaeus says these stars mark the raising of Jupiter by Themis and the nymph Amalthea, to whom his mother Ops entrusted him, and that Amalthea had a favorite goat which nursed Jupiter. Some, however, say it is Aega the daughter of Sol, who was renowned for the brightness of her body, but who, in contrast to her beautiful body, had an unsightly countenance. Because of this, the Titans, terrified, begged Terra to hide her body, and Terra was said to have hidden her in a cave on the island of Crete. Later Aega became the nurse of Jupiter, as we noted earlier. Now when Jupiter, with youthful confidence, was preparing for war against the Titans, the oracle told him that if he wished to prevail, he must be covered by a goat's skin and wage war with the head of the Gorgon—an article the Greeks call *aegis*. When he had done this, as we said earlier, Jupiter defeated the Titans and came to power. He covered the remaining bones of the goat with a goat's skin, brought the goat to life, and depicted her in a constellation, thus granting her immortality. Later, he gave to Minerva that which protected him when he was victorious.

Euhemerus said that a certain Aega ["goat"] was the wife of Pan and that, ravished by Jupiter, she gave birth to a son said to be Pan's. Thus, the boy was called Aegipan, and Jupiter was called *aegiochus* ["aegis-bearing"]. Because he loved Aega greatly, Zeus placed the figure of a goat among the stars, so that she would be remembered.

3.12

The figure has one star on the head; one on each shoulder (the one on the left shoulder, called Capra [Capella], is brighter); one on each elbow; one on the right hand; two on the (left) hand which are called Kids, located near the western stars. The total is eight.

Commentary

The accounts of Ps-Eratosthenes and Hyginus follow the same general outline: both describe the constellation figure as a charioteer; both identify the Goat as the animal that nursed the infant Zeus; and both provide an account of the origin of the aegis.

The charioteer depicted in this constellation is either Erichthonius, Myrtilus, or Orsilochus. Like most earth-born creatures of Greek mythology, Erichthonius is closely associated with serpents. Sometimes he is depicted as a serpent from the waist down, sometimes he is said to be the serpent behind the shield of Athena. Erichthonius was a great innovator according to ancient tradition. He introduced the chariot, or at least the four-horse chariot, and the *parabates*, i.e., the passenger who rode alongside the charioteer in some races. He either invented coinage or introduced the use of silver into Attica, and he instituted the Panathenaean Games.[1]

Myrtilus was the charioteer of Oenomaus, the king of Pisa and father of Hippodamia. Hippodamia persuaded Myrtilus to sabotage her father's chariot in the race against Pelops, and thus Myrtilus was responsible for Oenomaus's death. Before he died, Oenomaus cursed Myrtilus and prayed that he might die at the hands of Pelops. Myrtilus later tried to rape Hippodamia and was thrown into the sea by Pelops.[2]

Amalthea was either the nymph whose goat suckled the infant Zeus, or the goat herself. In the latter case, she was usually said to have nourished the infant Zeus with the nectar and ambrosia which flowed from her horns. In some accounts, Amalthea was one of three nymphs who reared Zeus on Crete. The three nymphs were usually said to be Cretan, but in a few accounts they were said to be Dodonian. According to Hyginus, while Cronus was searching for Zeus, Amalthea placed the infant in a cradle which she hung from the branch of a tree, so that Zeus was not to be found either in the sky or on land or in the sea.[3]

This constellation is probably of Babylonian origin. The figure of a charioteer standing in a four-horse chariot and accompanied by three goat-like animals appears on a Babylonian cylinder seal.[4]

Auriga is comprised of eight stars according to Ps-Eratosthenes, Hyginus, and Hipparchus, fourteen according to Ptolemy.

Bootes

The Constellations 8

This is said to be Arcas, the son of Callisto and Zeus. <Callisto lived around the sacred precinct of Zeus Lycaeus after Zeus violated her. Hesiod says that> Lycaon, <pretending not to know of the seduction>, cut up the infant and served him at the table when he entertained Zeus. The god, <in disgust at the king's inhumanity>, overturned the table, whence the city Trapezus received its name, then struck the house with lightning <and changed Lycaon into a wolf>. Arcas was created anew, made whole by Zeus, <and then raised by a goatherd>. It is said that as a young man Arcas entered the sacred precinct of Zeus Lycaeus and wed his mother, unaware of her true identity. When the inhabitants were about to sacrifice them both, in keeping with the law, Zeus carried them away because of his prior relationship with Callisto and placed them among the stars.

The figure has four stars on the right hand [ψ, 46, 45, ?]; these stars do not set. There is one bright star on the head [β]; one bright star on each shoulder [γ, δ]; one star on each breast [η CrB?, o CrB?], the one on the right breast being bright. Beneath this last star is one faint star [?]. There is one bright star on the right elbow [?]; one very bright star between the knees [α], which is called Arcturus; and one bright star on each foot [η, ζ]. The total is fourteen.

Poetic Astronomy 2.4

Concerning this constellation, it is said to be Arcas, the son of
Callisto and Jupiter, whom Lycaon reportedly cooked with other
meats, then placed before Jupiter as a meal when the latter sought his
hospitality, for Lycaon desired to know whether he who sought his
hospitality was a god. For this he brought on himself no small
punishment: Jupiter immediately overturned the table and destroyed
his house with a thunderbolt, changing him into a wolf; but the boy's
parts he collected and put back together and entrusted him to an
Aetolian to raise. The boy, now a youth, was hunting in the forests,
when, unknowingly, he encountered his mother who had been changed
into a bear. Intent on killing her, he chased her to the temple of
Lycaean Jupiter. Under Arcadian law, anyone entering this temple
incurred the penalty of death. And so, it was necessary to kill them
both, but Jupiter, pitying them, snatched the two away and placed
them among the stars, as we noted earlier [2.1]. Because of this, the
constellation appears to be following the Bear, and because it watches
the Bear, is called Arctophylax ["Bear-keeper"].

Many say that this is Icarius, the father of Erigone. To him, because
of his justice and piety, Liber granted wine, the vine, and the grape, so
that Icarius might show mankind in what way the vine should be
cultivated, what grows from it, and when it is grown, how it should be
used. When he had planted the vine and by diligent pruning made it
sprout, it is said that a goat fell on the vine and picked off all the tender
leaves it found there. Icarius became so irritated that he killed the goat,
made a bag out of its skin, filled the bag with air then tied it, and
throwing it among his companions ordered them to dance around it.
And so as Eratosthenes says, "Men first danced around the goat of
Icarius."

Others say that when Icarius received wine from Liber, he immediately loaded full wineskins upon a cart; and because of this was called
Bootes ["ox-driver"]. He traveled about introducing wine to the

shepherds in the region of Attica, many of whom, in their greed, took willingly to the new type of drink and becoming numb in their senses, talked of unseemly things and laid themselves down hither and yon, as if half-dead. The other shepherds, believing Icarius to have given their companions poison in order to drive their flocks away to his borders, killed Icarius and threw him into a well or, as others recount, buried him under a certain tree. When they awoke, however, the sleepers said they had never rested better and sought Icarius in order to thank him for his gift. Stung by their conscience, Icarius's slayers condemned themselves to exile and came to the island of the Ceans. There they were received as guests and built themselves homes.

Moved by longing for her father when she did not see him return, Icarius's daughter, Erigone, prepared to go in search of him. The dog of Icarius, whose name was Maera, returned to Erigone howling and appearing to lament its master's death. This appeared to Erigone to be a strong indication of death, for the timid maiden could not imagine anything other than that her father, who had been absent so many days and months, had been killed. Taking her robe in its teeth, the dog led her to the body. When the girl saw it she was overcome with despair, loneliness, and grief and, shedding many tears, resolved to hang herself from the very tree under which her father was buried; and the dog appeased the dead girl with its own spirit. Some, however, say the dog threw itself into the well which is called Anhigrum, and since that time no one has drunk from the well. Jupiter, pitying the three deaths, changed their bodies into stars. Thus many identify Bootes with Icarius and Virgo with Erigone (we will speak later about Virgo [2.25]), but they call the dog Canicula, because of its name and aspect. Canicula is called Procyon by the Greeks, because it rises before the greater Dog. Others say that maiden and dog were placed among the stars by Liber.

Meanwhile, many maidens in the Athenian land committed suicide by hanging for no apparent reason, for Erigone, dying, had prayed that the daughters of the Athenians should be afflicted with the same death she was about to suffer, until such time as the Athenians found the

murderer of Icarius and punished him. When the events we have
recounted occurred, the Athenians petitioned Apollo and his oracle
told them that if they wished to be freed of this matter, they must
appease Erigone. Because she hanged herself, they instituted the
custom of suspending themselves by ropes, with a plank attached, so
that the one hanging was moved by the wind. They celebrate this
sacrifice publicly and privately and call it Aletidae, and they call
Erigone "beggar"—*aletis*, in Greek—because she sought her father
with a dog, as was appropriate for one unknowing and alone. Now
Canicula, in rising, deprived the Cean lands and fields of crops with its
heat and afflicted the Ceans themselves with sickness, forcing them to
pay the penalty to Icarius with their own suffering, since they harbored
his murderers. Their king, Aristaeus, the son of Apollo and Cyrene
and father of Actaeon, inquired of his father by what act he might free
his city from calamity. The god ordered him to expiate the death of
Icarius with many sacrifices and to pray Jupiter that at the time
Canicula rises, he might send a wind for forty days that would
moderate the heat of Canicula. This command Aristaeus carried out,
and obtained an omen from Jupiter that the Etesian winds would blow.
Many call these winds *etesias*, because they arise at a certain time each
year, for the Greek word for year is *etos*; some call them *aetesias*,
because they were "requested" of Jupiter and granted. But that matter
we will leave unresolved, lest we appear to have thought of everything.

But to return to the matter at hand, Hermippus, who wrote about
the stars, says that Ceres mated with Iasion, the son of Electra; and
many, including Homer, say that because of that union Iasion was
struck by lightning. From that union, as the historian Petellides of
Cnossus recounts, two sons were born, Philomelus and Plutus, who
were said to be at odds with one another, because Plutus, who was
wealthier, gave none of his wealth to his brother. Pressed by circum-
stances, Philomelus took all he had and bought two oxen, then
invented the plough. And thus, by plowing and cultivating the fields,
he was able to feed himself. His mother, marveling at his inventions,
placed him among the stars as a ploughman, and called him Bootes.

They say he had a son, Parias, who named the Parians and the town of Parium after himself.

3.3

This figure has four stars on the right hand, which are said never to set; one on the head; one on each shoulder; one on each breast (the right one being brighter); and under it another, fainter, one; and one bright star on the right elbow. On the belt is one star brighter than the others, which is called Arcturus; and there is one star on each foot. The total is fourteen.

Commentary

Both Ps-Eratosthenes and Hyginus identify Bootes with Arcas, the son of Zeus and Callisto, and eponymous ancestor of the Arcadians.[1] The numerous traditions connecting Lycaon with human sacrifice may well be attempts to explain the rites performed in honor of Zeus Lycaeus on Mount Lycaeum, as Frazer suggests. According to Pausanias, the cult of Zeus Lycaeus and the Lycaean festival were instituted by Lycaon. Human sacrifice was associated with the cult from the beginning—Lycaon was said to have sacrificed a human infant—and there is literary evidence to suggest human sacrifice to Zeus Lycaeus down to the third century C.E.[2]

Ps-Eratosthenes cites Hesiod as his source for both the present story and the story of Ursa Major, below. The present story provides a logical connection with the story of Ursa Major, whether the two stories are taken as parts of the same narrative, as some suggest, or not.[3]

The name Arctophylax, mentioned by both Ps-Eratosthenes and Hyginus in connection with the story of Arcas and Callisto, derives from the connection of this constellation, in myth and position in the sky, with the constellations of the two Bears (Ursa Major and Ursa Minor). Homer refers to this constellation as Bootes, exclusively, and does not associate it with the two Bears, either in name or in myth. In recounting the stories of Icarius and Philomelus to explain why this constellation is named Bootes, Hyginus himself alludes neither to the name Arctophylax nor to any connection with the constellations Ursa Major and Ursa Minor. Indeed, in the narrative about Icarius, Bootes is associated with two entirely different constellations—Virgo and Canis Minor. It would appear, then, that the name Bootes is earlier than Arctophylax, and that the stories of Icarius and Philomelus, as constellation myths, may antedate the constellation myth of Arcas.[4]

The concept of a "guard" in connection with Bootes may reflect the constellation's origin in the Babylonian shepherd-constellation, which was believed to represent the shepherd of the "heavenly flock," i.e., the stars.[5]

Hyginus's narrative concerning Icarius contains not only two additional star myths (Virgo and Canis Minor), but also two aetiological stories explaining the origin of the Etesian winds and of the Athenian custom of swinging from trees during the festival of Anthesteria.[6] The Athenian festival of the Anthesteria entailed three days of ceremonies in honor of Dionysus. Among the practices observed was the swinging of maidens on swings suspended from trees.[7]

The Etesian winds provided relief from the debilitating heat of midsummer associated with the heliacal rising of Sirius (α CMa), which was believed to cause sickness and weakness in men, wantonness in women, and the parching of crops.[8] Hyginus associates the midsummer heat with Procyon (α CMi), which he says is also called Canicula ("little dog").[9]

The number of stars in Bootes is fourteen according to Ps-Eratosthenes and Hyginus, nineteen according to Hipparchus, and twenty-two according to Ptolemy.

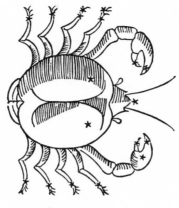

Cancer

The Constellations 11

The Crab [Cancer] is thought to have been placed among the stars by Hera, because, when all the other animals favored Heracles in his struggle against the hydra, the crab jumped out of the lake and bit him on the foot, as Panyassis recounts in his *Heraclea*. Angered, Heracles allegedly crushed the crab with his foot. For this reason, the Crab received great honor, being numbered among the twelve signs of the zodiac.

Some of the stars in this constellation are called the Asses [Asini]. These were placed among the stars by Dionysus. Their distinguishing sign is the Manger [Praesepium], and their story is the following.

When the gods were attacking the Giants, it is said that Dionysus, Hephaestus, and the Satyrs rode [to battle] on asses. As they approached the Giants, who were not yet visible, the asses brayed, and the Giants, hearing the noise, fled. For this reason the asses were honored, being placed on the western side of the Crab.

The Crab has two bright stars on its shell [γ, δ]; these are the Asses. The nebula visible within the Crab is the so-called Manger [ε], beside which the Asses are standing. There is one faint star on each of the four right legs of the Crab [β, π?, κ, ?]; two faint stars on the first of the left

61

legs [μ, ?]; two stars on the second of the left legs [?, ?]; one on the third [?]; and one on the edge of the fourth [?]. In addition, there is one star on the mouth [?]; three on the right claw [α, κ?, π?], similar to one another and small; and two on the left claw, also small [ι, ?]. The total is eighteen.

Poetic Astronomy 2.23

This constellation was placed among the stars by Juno, because when Hercules was struggling with the Lernaean hydra, the crab came out of the swamp and seized his foot with its teeth. Hercules, greatly angered, killed the crab, but Juno placed it among the stars, to be one of the twelve signs which are circumscribed by the sun's course.

In one part of this figure are certain stars called Asses, depicted on the shell of the Crab by Liber, and consisting of only two stars. Liber, suffering from a madness caused by Juno, was said to have fled through Thesprotia, hoping to reach the oracle of Jupiter at Dodona, where he could inquire how he might most easily recover his sanity. Coming to a large impassable swamp, he met two asses, captured one of them, and so crossed the swamp without coming into contact with the water. When he arrived at the temple of Jupiter at Dodona, he was said to have been cured instantly of his madness. He thanked the asses by placing them among the stars.

Some say that he bestowed a human voice on the ass which carried him. Later, when that ass was defeated by Priapus in a dispute over his male member, Priapus killed the ass, but Liber, in pity, placed the ass among the stars. And that it might be known he did this as a god, and not as a fearful mortal fleeing Juno, Liber placed the ass above the Crab, which had been placed among the stars by that goddess.

Another story about the Asses is told by Eratosthenes. When Jupiter waged war against the Giants and summoned all the gods to fight against them, Liber, Vulcan, Silenus and the Satyrs came on asses. As they approached the Giants, the asses became frightened and made a great noise the Giants had never heard before; on account of that noise the Giants fled and thus were defeated. Similar to this story is the story about Triton's shell, for, when he hollowed out the shell-horn he had invented, he, too, came to fight the Giants and made an unheard of noise with his shell. The Giants, fearing that the enemy had unleashed a great beast whose sound they were hearing, fled and, defeated, came under the power of their enemies.

3.22

The figure has two stars on the head which are called Asses, about which we spoke earlier; a faint star on each right foot; two on the foremost left foot; two on the second; one on the third; one faint star on the fourth; one on the mouth; three similar, small stars on what is called the right claw; two like stars on the left claw. There are eighteen stars in all.

Commentary

Both Ps-Eratosthenes and Hyginus identify the figure in this constellation with the crab that attacked Heracles as he was slaying the Lernaean hydra, and the two stars named Asini with the animals that carried Dionysus, Hephaestus, and the Satyrs to battle, and were responsible for the victory of the Olympian gods against the Giants. Hyginus provides an additional story linking Asini with Dionysus.[1]

The slaying of the Lernaean hydra, the second of the Labors of Heracles, is referred to by numerous authors, but there is only one reference aside from that in *The Constellations* to the crab sent by Hera. Hera's antagonism toward Heracles is typical of her attitude toward all Zeus's paramours and their offspring.[2]

Of the two additional figures within this constellation, the pair of stars known as Asini (γ and δ Cnc) are associated with Dionysus. The connection of Dionysus with bulls, goats and, to a lesser degree, donkeys, all animals symbolizing fertility and sexual drive, is well known from ancient art as well as literature. Hephaestus is often associated with Dionysus, and the Satyrs are usually close companions of Dionysus. As for the figure of the Manger, no classical source speaks of its origin as a constellation. The story of Dionysus and the asses may represent a very old tradition connected with this constellation. Some eastern zodiacs depict the constellation now known as Cancer as two asses.[3]

Artistic representations of a crab constellation appear on Babylonian cylinder seals and tablets, but whether the Babylonian constellation corresponded exactly to the Greek constellation is uncertain. The Greek constellation can be traced as far back as Eudoxus.

The number of stars in Cancer is eighteen according to Ps-Eratosthenes, Hyginus, and Hipparchus, nine according to Ptolemy.

Canis Major

The Constellations 33

   This figure is said to be the dog given to Europa as guard, along with a spear. Minos later presented both dog and spear to Procris, after she cured him of an illness. Soon after, Cephalus gained possession of dog and spear by marrying Procris. Cephalus brought the dog to Thebes to chase the fox which, it was said, no one could slay. Caught in a dilemma, Zeus turned the fox to stone and placed the dog among the stars, judging it worthy of the honor. Others say this constellation is the dog of Orion which followed him on the hunt, since all hunters believe that a dog helps to ward off wild animals. They say that the dog was placed among the stars when Orion was changed into a constellation, and that this was done, in all likelihood, so that Orion would lack none of his belongings.

   The figure has one star, which is called Isis or Sirius, on the head [α]. This star is large and very bright and stars similar to it are called sirii by astronomers because of the scintillation of their light. There is one bright star, called Cyon ["dog"], on the tongue [α]. There are two stars on the neck [γ, ι]; one faint star on either shoulder [o¹?, o²?]; two stars on the chest [π, ?]; three on the front foot [β, ν³?, ν²?]; three on the back [δ, ?, ?]; two on the belly [ε, ?]; one on the left haunch [δ]; one on the

65

edge of the paw [ζ]; one on the right paw [κ]; and one on the tail [η]. The total is twenty.

Poetic Astronomy 2.35

This dog was reportedly appointed by Jupiter to guard Europa; thus he came into the possession of Minos. When the latter was ill, he was cured by Procris, the wife of Cephalus, who received the dog as a reward for the service rendered, for she was very fond of hunting and the dog had the power to outrun any wild beast. After her death, the dog came into the possession of Cephalus, since Procris was his wife, and he brought the dog to Thebes, where there was a fox said to be capable of outrunning any dog. When dog and fox came together, Jupiter, in a quandary, changed both to stone, as Istrus recounts.

Many say that this is the dog of Orion and that because Orion was a zealous hunter, the dog was placed with him among the stars. Others, however, say it is the dog of Icarius, about whom we spoke earlier [2.4]; and many other ideas are put forth. Canis ["dog"] has one star on the tongue which is also called Canis, and another on the head which Isis herself reportedly placed there and called Sirius on account of the brightness of its light, which is such that it outshines all other stars. She called the star Sirius ["scintillating"] so that it might be recognized more easily.

3.34

The figure has one bright star, called Canis, on the tongue; another on the head, which some call Sirius, and about which we spoke above. In addition, there is one faint star on each ear; two on the chest; three on the forward foot; three on the back; one on the left haunch; one on the rear foot; one on the right foot; four on the tail. The total is nineteen.

Commentary

Beginning with Odysseus's faithful dog, which endured its master's twenty-year absence only to expire on his return, particular dogs are usually represented in Greek and Latin literature as loyal companions of mankind.[1] Such are the creatures identified by Ps-Eratosthenes and Hyginus in connection with Canis Major. This constellation is said to represent the dog of Procris, Orion, or Icarius—the third possibility being posed only by Hyginus.[2]

The Constellations reflects the earliest reference to the figure in Canis Major as the dog of Procris. The fox of the story is the beast sent by the gods to plague the Thebans because they had barred the descendants of Cadmus from the kingship. The Thebans were obliged to expose the son of a citizen to the fox each month. Finally, they asked Cephalus to bring his dog to Thebes and to rid the city of the fox. Since both dog and fox were invincible,[3] Zeus brought an end to the chase by changing one or both of them to stone.

The earliest Greek reference to this constellation as Orion's dog is in Homer, although the reference may be to Sirius, rather than to the entire constellation.[4] There are no stories concerning Orion's dog independent of its connection with this constellation.

Sirius was the brightest of all stars in antiquity, as it is today; nevertheless, it was of no particular significance in Greek mythology. In Greek literature, the rising of Sirius is connected with the scorching heat of midsummer.[5]

The non-Greek origin of this constellation is suggested by the association of both the constellation and its brightest star with a dog in Euphratean and Egyptian records, and from reference in Euphratean records to the "heavenly hunter and his dogs."[6]

The number of stars in Canis Major is twenty according to Ps-Eratosthenes and Hipparchus, eighteen according to Ptolemy.

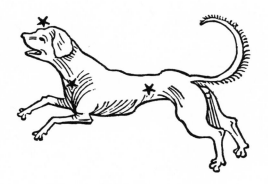

Canis Minor

The Constellations 42

This figure lies in front of the Great Dog [Canis Major] and is called Procyon ["Dog Preceder."] It is said that Procyon was placed next to Orion because of the latter's love of hunting. Indeed, the Hare follows Orion and other wild animals are visible beside him.

The figure has three stars [α, β, ?], of which the first is bright when it rises. This star bears a resemblance to the [Great] Dog; for this reason it is called Procyon and rises and sets before the Great Dog.

Poetic Astronomy 2.36

This constellation is seen to rise before the Great Dog. Many consider it to be the dog of Orion, for which reason it is called Procyon. The stories we have recounted above about Canis Major apply to him as well.

Commentary

The name of Canis Minor and the myths connected with it point to its close association—and, often, confusion—in antiquity with the constellation Canis Major. The name Procyon was given to this constellation either because of its position "in front of" Canis Major or because it rises before the latter. The constellation Procyon had no mythology of its own among the Greeks. The two identifications of the dog represented in Procyon are applied also to Canis Major by Hyginus: Procyon is either Orion's dog (in which case Canis Major is to be identified with some other dog), or Icarius's dog, but there is some uncertainty involved in the latter identification.[1]

As in the case of Cyon, the name Procyon was used by the Greeks for both the constellation and its brightest star. The constellation appears to be of late origin, since Procyon is still regarded as a star (α CMi) by Aratus. Euphratean as well as Egyptian records refer to the star α CMi in connection with a dog, but the constellation Canis Minor may be of Greek origin.[2]

The number of stars in this constellation is three according to Ps-Eratosthenes, Hyginus, and Hipparchus, two according to Ptolemy.

Capricorn

The Constellations 27

Aegocerus ["Goat-horn"] is similar in appearance to Aegipan, and is descended from him. His lower members are those of a wild animal and he has horns on his head. According to Epimenides, the author of the *Cretica*, Aegocerus was honored because he was raised together with Zeus and because he accompanied Zeus when the latter fought against the Titans on Mount Ida. Aegocerus is thought to have invented the trumpet which is called Panicus ["panic"] on account of its sound; for as Aegocerus was arming the allies, the sound of his trumpet caused the Titans to flee. After Zeus assumed power, he placed Aegocerus among the stars along with his mother, Aega ("Goat"). Aegocerus has a fish's tail as his attribute because he discovered the trumpet in the sea.

Aegocerus has one star on each horn [$α^{1,2}$, $ξ^{1,2}$]; one bright star on the nose [o?]; two stars on the head [π, ρ]; one beneath the neck [υ?]; two on the chest [24, ?]; one on the front leg [?]; one at the edge of the foot [?]; seven on the back [θ, ι, ϵ, κ, γ, δ, ?]; five on the belly [φ, χ, η, ζ, 36]; two bright stars on the tail [42, λ?]. The total is twenty-four.

71

Poetic Astronomy 2.28

Capricorn's ["Goat-horn"] appearance is similar to that of Aegipan. Because he was raised together with Jupiter, the latter wished Capricorn to be among the stars, along with their nurse, Capra [Capella], of whom we spoke earlier [2.13]. Eratosthenes recounts that when Jupiter was fighting against the Titans, Capricorn first inspired in the enemy that fear which is called *panikos* ["panic"]. Because of this, and because he showered the enemy with murex shells instead of stones, his lower body is in the form of a fish.

Egyptian priests and some poets, however, say that when many of the gods gathered in Egypt, they were suddenly set upon by Typhon, a horrible Giant and great enemy of the gods. The gods, greatly frightened, changed themselves into different shapes: Mercury into an ibis, Apollo into a raven, Diana into a cat. For that reason, these animals are held to be inviolate by the Egyptians, because they are considered to be images of the gods. On that same occasion, Pan reportedly jumped into the river, changed his hind parts into a fish, and the rest of his body into a goat, and thus escaped from Typhon. Jupiter, admiring Pan's ruse, placed that image among the stars.

Capricorn has one star on each horn; one on the nostril; two on the head; one under the neck; two on the chest; one on the front foot; one on the end of the same foot; seven on the back; five on the belly, two on the tail. There are twenty-four stars in all.

Commentary

It is uncertain whether Aegipan is to be considered as a doublet of Pan or as a separate being. Pan himself appears in classical literature mostly as a human figure, but with goat's legs and horns. This image of Pan persists in classical art; however, there are some representations of Pan in entirely human form. The name Aegipan ("Goat-pan") may refer to the hybrid form of Pan.[1]

The Constellations describes Capricorn as having both the lower members of a wild animal, and a fish's tail. This apparent contradiction may reflect an attempt on the part of Ps-Eratosthenes to reconcile the traditional Greek image of Aegipan with the Babylonian constellation-figure corresponding to Capricorn, which depicts a hybrid creature whose upper half is a goat while its lower half is a fish's tail.[2]

Capricorn appears as a distinct entity only in astronomical literature. Although *The Constellations* alludes to Capricorn's presence on Mount Ida during the infancy of Zeus and the Titanomachia, Greek myth usually limits the companions of the infant Zeus on Mount Ida to the attendants of the god: the nurse Amaltheia, the Curetes, and various other nymphs. Capricorn is mentioned as a part of this group of attendants only by *The Constellations*, Hyginus, and related astronomical literature.[3]

The mention of Mount Ida as the site of the Titanomachia is also unique to *The Constellations*. The traditional site of the battle between the Titans and the Olympian gods is Thessaly. However, as ancient authors were wont to do, the author of *The Constellations* may have confused the Titanomachia with the Gigantomachia. The latter is set in various locations by ancient sources.[4]

The sound of Pan's shell-trumpet was believed to cause panic—the often groundless fear which possesses human beings in large groups. The earliest reference to use of the shell-trumpet in the Titanomachia or Gigantomachia is in *The Constellations*. A similar shell-trumpet is a normal attribute of Triton who, however, used its sound not as a means of inspiring fear, but to calm the sea.[5]

Both Ps-Eratosthenes and Hyginus identify this figure as "Goat-horn" [Capricorn]. A late artistic representation of the constellations shows a Nereid in place of Capricorn, but this may reflect an attempt to replace the Babylonian constellation-figure, which was closely connected with water, by a Greek water-spirit.[6]

The number of stars in Capricorn is twenty-four according to Ps-Eratosthenes and Hyginus, twenty-six according to Hipparchus, and twenty-eight according to Ptolemy.

Cassiopeia

The Constellations 16

Sophocles the tragedian says in his *Andromeda* that Cassiopeia was punished because she challenged the beauty of the Nereids. Poseidon sent a sea monster to ravage the land and Andromeda was exposed to the monster because of her mother. Accordingly, Cassiopeia is represented in the sky as seated on a chair, close to Andromeda.

Cassiopeia has one bright star on her head [ζ]; one bright star on each shoulder [?, ?]; one bright star on the right breast [α]; one star on the elbow [φ]; one bright star on the right hand [σ?]; one bright star on the left hand [θ]; one star on the navel [η?]; one on the knee [?]; one on the edge of the foot [ι]; one faint star on the head [?]; two bright stars on the left thigh [γ, ?]; one bright star on the knee [δ]; one star on the edge of the foot [?]; one on the seat of the chair [κ]; one on either corner of the chair on which she is seated [β, ρ]. The total is fifteen.

Poetic Astronomy 2.10

Concerning Cassiopeia, Euripides and Sophocles, along with many others, say she boasted that her beauty surpassed that of the

Nereids. Because of this she was placed among the stars, seated on a throne[?], and because of her impiety, she appears to be borne head downward as the sky revolves.

3.9

She has one star on the head; one on each shoulder; one bright star on the right breast [. . .] one large star on the navel; two on the left thigh; one on the knee; one on the forward right foot; one in the square which forms the seat of the chair; one on each corner, shining more brightly than the others. The total is fourteen.

Commentary

The story of mortals who, in their arrogance, challenge the gods in skill or beauty and bring disastrous consequences upon themselves and their families is a common one in Greek mythology. Medusa boasted that her beauty equaled that of Athena and was turned into a terrifying monster; Agamemnon boasted of his prowess as a hunter and was made to offer his daughter as a sacrificial victim. Cassiopeia was considerably more fortunate than most mortals who challenged the divine powers. Poseidon avenged the insult to the Nereids by sending a sea monster to Cepheus and causing Andromeda to be exposed to the creature, but the monster was soon dispatched by Perseus, and Cassiopeia, along with the other principals of the story, was changed into a constellation. Hyginus interprets the placement of Cassiopeia in a head-downward position as punishment for her hybris.[1]

The constellation of Cassiopeia appears to be of Phoenician origin, Cassiopeia herself representing the counterpart of the king-figure in the constellation Cepheus.[2]

The Nereids were sea-nymphs and daughters of Nereus and Doris. Ninety-eight names of sea-nymphs are mentioned in classical literature. A few individual Nereids figure in Greek myths: Thetis, the mother of Achilles; Amphitrite, the wife of Poseidon; Oreithyia, the mother of Calais and Zetes; and Galateia, the beloved of Polyphemus. The Nereids were kindly spirits, especially helpful to mariners in distress.[3]

Cassiopeia is comprised of fifteen stars according to Ps-Eratosthenes, fourteen according to Hyginus, thirteen according to Ptolemy.

Centaurus

The Constellations 40

This figure is believed to be Chiron, who inhabited Mount Pelion and surpassed all men in righteousness. He was the tutor of Asclepius and Achilles. Heracles was believed to have approached him out of love, and Chiron, honoring the god Pan, reportedly had intercourse with him in the cave. Chiron was the only one of the Centaurs that Heracles did not slay, but conversed with, as Antisthenes the Socratic recounts in his *Heracles*. As they engaged in long conversation, an arrow fell out of Heracles's quiver and onto Chiron's foot, causing his death. Zeus, on account of Chiron's piety and the misfortune that befell him placed him among the stars. The Wild Animal [Lupus], which he is believed to be about to sacrifice, is in his hands, close to the Altar [Ara]—a very great manifestation of his piety.

He has three faint stars above his head [2, 4?, 1?]; one bright one on each shoulder [ι, θ]; one on the left elbow [χ?]; one on the tip of his hand [κ]; one in the middle of his equine chest [M]; one on each front hoof [α, β]; four on his back [ζ, υ², υ¹, ω]; two bright stars on the belly [ε, Q]; three on the tail [?, ?, ?]; a bright star on his equine haunch [γ]; one on each back knee [γ Cru, δ Cru]; one on each hoof [α Cru, μ Cru]. There are twenty-four in all. In his hands he holds the so-called Wild

Animal [Lupus], which has the shape of a square. Some say this is a wine-skin, out of which he is pouring a libation on the Altar [Ara], in honor of the gods. He holds this in the right hand, and a thyrsus in the left hand [ψ, c¹]. The Wild Animal has two stars on its tail [ι Lup?, ρ Lup?]; one bright one on the edge of the back foot [β Lup]; one between the feet [α Lup?]; one bright one on the neck [η Lup?]; one bright one on the front foot [2 Lup]; one under that foot [1 Lup]; three on the head [θ Lup?, φ¹,² Lup?, χ Lup]. There are ten in all.

Poetic Astronomy 2.38

This is said to be Chiron, the son of Saturn and Philyra, who surpassed not only the other centaurs but also men in justice. He was said to have tutored Asclepius and Achilles. He achieved a place among the stars on account of his piety and diligence. Once, when Hercules was Chiron's guest, as they sat together looking at arrows, one of the arrows fell on Chiron's foot and killed him.

Others, however, say that Centaurus expressed amazement that Hercules could fell the great bodies of the centaurs with such short arrows, and himself tried to shoot one with his bow. The arrow fell out of his hand and onto his foot. Because of this Jupiter pitied him and placed him among the stars with an offering which he appears to be sacrificing on the Altar. Others say this is the centaur Pholus, who was more expert by far than the others in divination. Thus, by the will of Jupiter, he was depicted as coming to the Altar with an offering.

3.37

He has three faint stars on his head; one bright one on each shoulder; one on the left elbow; one on the hand; one in the middle of the horse's chest; one on each front leg; four on the back; two bright ones on the belly; three on the tail; one on the equine haunch; one on each rear knee and one on the hollow of each knee. The total is twenty-four. The sacrificial offering has two stars on the tail; one on the forward rear foot; one between the feet; one bright one on the back; one in the front part of the feet; another above the feet; three on the head. The total is ten.

Commentary

The preeminent position of Chiron among the centaurs by virtue of his wisdom and skill is attested in Greek literature from the very beginning. Numerous Greek heroes were brought to Chiron's cave on Mount Pelion to be instructed in the arts of healing and hunting, in music, war and justice. Among his more famous pupils were Achilles, Asclepius, Jason, Heracles, Cephalus, Nestor, Meleager, Theseus, Odysseus, Diomedes, Castor and Polydeuces. Chiron was not only the teacher, but also the friend and counselor of his pupils; there are several stories illustrating Chiron's friendliness and service toward them.[1]

Chiron is usually said to be the son of Cronus and Philyra; however, some sources call him the son of a Naiad, others of Poseidon or of Ixion. His wife was either a Naiad or Chariclo, and his children are variously reported as Carystus, Aristaeus, Endeis, Thetis, Ocyrrhoe, Melanippe or Evippe.[2]

Chiron was immortal but willingly surrendered his immortality when he received an incurable wound from one of the poisoned arrows of Heracles. There are two accounts of Chiron's death. In one account, Heracles and Chiron were engrossed in conversation when an arrow slipped from Heracles's quiver, striking Chiron on the foot. In the other account, Heracles pursued the centaurs from Pelion to Malea, where they sought refuge with Chiron. He then began to shoot at them and one of his arrows struck Chiron on the knee. In both accounts Heracles tried to heal the wound but could not, and Chiron, unable to bear the pain, exchanged his own immortality for Prometheus's mortality, and so died.[3]

The form of Chiron described in *The Constellations* is that of a horse with a man's body from the waist up. While this form is in keeping with

the representations of other centaurs in art from the fifth century
B.C.E. on, it is somewhat unusual for Chiron, who tends to persist in
the earlier form of a complete man with the barrel and hind legs of a
horse attached to his back.[4]

Most ancient references to this constellation identify it with Chiron.
Hyginus states that some ancients believed the centaur represented
was not Chiron but Pholus. A late reference calls the constellation
Hippocrator ("ruler of horses"). It has been suggested that this
constellation is of Greek origin and represents a reduplication of the
constellation of the Archer (Sagittarius).[5]

The Wild Animal in the hand of the Centaur is the constellation
known today as Lupus. *The Constellations* provides no myth explaining
the origin of the Wild Animal, but other sources identify it with the
wolf into which Lycaon was changed. The constellation Lupus itself
is of Euphratean origin.[6]

The thyrsus, a wand wreathed with ivy and vine-leaves carried by the
worshippers of Dionysus, does not appear to have been connected
with the Euphratean constellation corresponding to the Archer. The
thyrsus became associated with the centaurs in the Graeco-Roman
period. The evidence for that association is from art exclusively and
includes a wall-painting and silver cup found at Pompeii.[7]

The number of stars in the Centaur is twenty-four according to Ps-
Eratosthenes and Hyginus, twenty-six according to Hipparchus and
thirty-seven according to Ptolemy.

Cepheus

The Constellations 15

Cepheus is the fourth in order of the constellations. From feet to chest he lies within the Arctic Circle. The remainder of his body lies between the Arctic Circle and the Tropic of Cancer. According to Euripides, Cepheus was king of the Ethiopians and the father of Andromeda. He is said to have exposed his daughter to a sea monster, but she was rescued by Perseus, the son of Zeus. Because of her, Cepheus, too, was placed among the stars, at the wish of Athena.

Cepheus has two bright stars on the head [δ?, ε?], one star on either shoulder [α, ?], and one on either hand [ι, ?], one faint star on either elbow [η, θ], three faint stars crosswise on the belt [β, ?, ?], at the waist, one star on the right hip [κ], two on the left knee [γ, ?], four above the feet [?, ?, ?, ?], one at the edge of the foot [?]. The total is nineteen.

Poetic Astronomy 2.9

According to Euripides and others, this figure is the son of Phoenix, king of the Ethiopians, and father of Andromeda. As related in well-known accounts, Andromeda was exposed by her father to the sea monster. She was rescued by Perseus, who took her as his bride. And so, that their entire race should enjoy immortality, the gods placed

Cepheus, too, among the stars.

3.8

The figure has two stars on the head; one on the right hand; one faint star on the elbow; two on the left hand and shoulder; one on the right shoulder; three bright stars on the belt which divides his body; one faint star on the right side; two on the left knee; one on each foot; four stars above the feet. The total number of stars is nineteen.

Commentary

Most of the figures appearing in Greek constellations were placed there by one of the gods to honor and perpetuate the memory of some notable personage or deed. The constellation-figures of Cepheus and Cassiopeia are unusual in that they were not granted as an honor, but to complement the constellations of Perseus, Andromeda, and Cetus. This group of five constellations is unusual in another respect. The five constellations represent all the principal figures of the Perseus-Andromeda myth. This is the only classical myth to be depicted so fully among the constellations.[1]

Cepheus appears in Greek literature as a foreign king, the husband of Cassiopeia and father of Andromeda. The hubris of Cassiopeia in claiming that her beauty surpassed that of the Nereids brought divine wrath upon her family.[2]

This constellation appears to be of Phoenician origin. At any rate, it does not appear on Babylonian cylinder seals or monuments, and ancient authorities attest that it was not known either to the Chaldeans or the Egyptians. Cepheus is the only personage identified with this constellation in classical literature.[3]

The number of stars in Cepheus is nineteen according to Ps-Eratosthenes, Hyginus, and Hipparchus, eleven according to Ptolemy.

Cetus

The Constellations 36

This is the beast that Poseidon sent to plague Cepheus, when Cassiopeia claimed to rival the Nereids in beauty. Perseus slew this sea-monster, and it was placed among the stars to commemorate his deed. Sophocles the tragedian tells the story in his *Andromeda*.

Cetus has two bright stars on the tail [ι, β]; five between the tail and the curvature of the flank [θ, η, τ, υ, ζ]; and six under the belly [ρ, σ, ε, π, ?, ?]. The total is thirteen.

Poetic Astronomy 2.31

Concerning Cetus, it is said he was sent by Neptune to devour Andromeda, about whom we spoke earlier [2.11]. He was slain by Perseus, and placed among the stars on account of the monstrous size of his body and Perseus's courage.

3.30

The figure has two faint stars at the end of the tail; five from there to the curve of its body; six under the belly. The total is thirteen.

Commentary

Sea-monsters are not common in Greek mythology. The sea-monster of the present story is not described by either Ps-Eratosthenes or Hyginus. However, artistic representations of this constellation show either a creature resembling a sea dragon or a creature with the body of a fish and the head of a dog. A sea dragon, rather than the hybrid dog-fish creature, is described by Aratus and subsequent astronomical literature.[1]

Greek, Egyptian and Babylonian origins have been suggested for this constellation. Those who advocate a Greek origin for Cetus suggest that the four constellations of Andromeda, Cassiopeia, Cepheus and Cetus were named in the late fifth or early fourth century B.C.E. by a learned individual, who grouped them about the already-known celestial figure of Perseus. The argument for the Babylonian origin of Cetus is purely speculative; the Egyptian origin for this constellation is advocated on the basis of the identification of Cetus with the Egyptian zodiacal constellation of the Crocodile.[2]

The number of stars comprising Cetus is thirteen according to Ps-Eratosthenes and Hyginus; fourteen, thirteen or eight according to Hipparchus, and twenty-two according to Ptolemy.

Corona Borealis

The Constellations 5

This is said to be the crown of Ariadne, which Dionysus, <wishing to manifest himself to the gods>, placed among the stars when the gods came to celebrate the marriage of Theseus and Ariadne at Dia. The bride received the crown from the Horae and Aphrodite and was the first to be crowned with it. <The author of the *Cretica* says that Dionysus presented the crown to Ariadne when he came to the realm of Minos with the intent of seducing her, and that Ariadne was tricked by means of it.>

The crown was said to be the work of Hephaestus and to be wrought of fiery gold and Indian gems. It is recounted that the light given off by this crown guided Theseus out of the labyrinth. <Later, after Theseus and Ariadne came to Naxos, the crown was placed among the stars, with the consent of the gods, as a symbol of the couple's success.> Ariadne's lock is said to be visible beneath the tail of the Lion [Leo].

The figure consists of nine stars lying in a circle [α, β, γ, δ, ε, θ, ι, π, ?]. Three of these stars are bright and lie opposite the head of the Dragon [Draco], which lies between the Bears [Ursa Major, Ursa Minor].

Poetic Astronomy 2.5

This is said to be the crown of Ariadne, placed among the stars by Liber. For it is recounted that at the marriage of Ariadne and Liber on the island of Dia, when all the gods brought gifts, Venus and the Horae ["Seasons"] presented the first [bridal] crown to Ariadne.

But the author of the *Cretica* writes that when Liber came to the realm of Minos with the intention of seducing Ariadne, he gave her this crown as a gift and she was so delighted by it that she did not refuse him. It was said to be the work of Vulcan, wrought of gold and Indian gems. With it, Theseus reportedly emerged from the darkness of the labyrinth, because the gold and gems produced a glow in the darkness.

The authors of the *Argolica* give the following explanation. Having received permission from his father to bring his mother, Semele, back from the Underworld, Liber came to the territory of the Argives, seeking the way to the Underworld. There he met a man called Polymnus—a man worthy of this generation—of whom he would inquire and who would show him the way. Polymnus, when he saw the youth, surpassing all others in physical beauty, asked of him a price which could be given without injury. Liber, eager to see his mother, swore that if he brought her back, he would do whatever Polymnus wished; and so, Polymnus showed him the way—but Liber swore as a god would swear to an impudent mortal. When Liber came to the place indicated and was about to descend, he placed the crown given to him by Venus in the place thereafter called Stephanus ["crown,"] for he did not wish to carry it with him, lest a gift of the gods should become polluted by contact with the dead. When he brought his mother back safely, he was said to have placed the crown among the stars so that its name might be remembered forever.

Others say this was the crown of Theseus and was placed there because the constellation called The Kneeler [Hercules] is believed to be Theseus, about whom we will say more later [2.6]. It was said that when Theseus came to Minos in Crete with the seven maidens and six youths, Minos, struck by the beauty of the maiden called Eriboea, would have raped her. Theseus, as a son of Neptune capable of contending against a tyrant for the sake of a maiden's well-being, said he would not countenance this. And so, the argument turned away from the girl and to the parentage of Theseus, whether or not he was the son of Neptune. Removing a small gold ring from his finger and throwing it into the sea, Minos challenged Theseus to recover it, to prove he was the son of Neptune—for he himself could easily prove he was the son of Jupiter. Minos prayed to his father, seeking some sign that he was his son, and immediately the sky responded with thunder and lightning. In his turn, Theseus, without any prayer or oath, threw himself into the sea; immediately a great horde of dolphins arose from the sea and brought him with gentle waves to the Nereids, from whom he recovered the ring of Minos, while Thetis gave him the gem-encrusted crown which she had received from Venus as a wedding gift.

Others say that Theseus received the crown from the wife of Neptune, and that he presented it to Ariadne as a gift when Ariadne was given to him as his wife on account of his courage and greatness of soul. After Ariadne's death, Liber placed the crown among the stars.

3.4

The constellation has eight stars arranged in a circle; of these, three shine more brightly than the others.

Commentary

Both Ps-Eratosthenes and Hyginus associate this constellation almost exclusively with figures from the Cretan cycle of legends: King Minos, his daughter Ariadne, her suitors Theseus and Dionysus, the Labyrinth; only the story of Liber's descent to the Underworld, recounted by Hyginus, departs from that pattern. The structure of the central narrative is a familiar one: a father (King Minos) puts his daughter's suitor (Theseus) to the test (confronting the Minotaur in the Labyrinth); the suitor succeeds thanks to the possession of an object, usually provided by the king's daughter. In the traditional account of the legend, Theseus found his way out of the Labyrinth by rewinding a ball of string provided by Ariadne, which he attached to the entrance. In the accounts of Ps-Eratosthenes and Hyginus, uniquely, the stratagem of the string is replaced by a crown that glowed in the dark, acquired by Theseus from either Thetis or Amphitrite.[1]

According to one tradition, the constellation Corona Borealis represents a jewel-studded crown usually associated with Ariadne. The crown itself was fashioned by Aphrodite's husband, Hephaestus, blacksmith to the Olympian gods, and was presented by the goddess to one or more brides (Ariadne, Thetis), or to Dionysus. Another tradition identifies the constellation with the ivy wreath of Dionysus.[2]

Ariadne herself probably represents a version of the Cretan mother-earth goddess. She was the daughter of Minos and Pasiphae ("all-shining")—herself a mother-earth goddess. In addition, one of Ariadne's Cretan names was Aridela ("bright"). Ariadne is closely connected with Aphrodite and was worshipped on the islands of Naxos and Cyprus as Aphrodite-Ariadne. The tradition that Ariadne was the mother of Tauropolis suggests a connection with Artemis Tauropolus, another mother-earth goddess. There was also a cult of Ariadne in Attica.[3]

The identification of the island of Dia, where the wedding of Ariadne and Theseus, or of Ariadne and Dionysus, was celebrated, cannot be determined with absolute certainty, since several islands bore that name in antiquity. The most likely choice is either the present-day Naxos, or the island now called Standia, off the north coast of Crete, since both islands were sacred to Dionysus in ancient times.[4]

After coming to the island of Dia/Naxos, Ariadne was either slain by Artemis—with the approval of Dionysus, whereupon Theseus, overcome with grief, sailed away— or she was abandoned by Theseus and became the wife of Dionysus. That Ariadne died on Naxos is stated explicitly or implied by most ancient authorities. The wedding of Ariadne and Theseus or Dionysus is referred to by numerous ancient authors. Ps-Eratosthenes's mention of the fact that Ariadne was the first bride to wear a crown is an aetion for the use of the crown in the ancient (and modern) Greek marriage rite.[5]

The descent of Liber to the Underworld to recover his mother, Semele, is reminiscent of the descent of Orpheus to recover his wife, Eurydice. Orpheus was closely associated with Dionysus: not only did he introduce the cult of Dionysus wherever he went, he also met his death in the course of a Dionysiac ritual.[6]

The number of stars in Corona Borealis is variously given as nine (Ps-Eratosthenes), eight (Hyginus and Ptolemy), and five (Hipparchus).[7]

Cygnus

The Constellations 25

This constellation [Cygnus] is called the Great Bird and is believed to be a swan. It is said that Zeus, enamored of Nemesis, transformed himself into this bird when she changed her shape and assumed the form of a swan in order to protect her virginity. Zeus changed himself into a swan and flew down to Rhamnus in Attica, where he ravished her. She bore an egg, from which Helen was hatched, as the poet Cratinus recounts. Because he did not change his shape but flew away thus to heaven, Zeus placed the image of a swan among the stars. The swan is represented as it was at that time, in flight.

The figure has one bright star on the head [β]; one bright star on the neck [η]; five stars on the right wing [δ, θ, ι, κ, ?]; five on the left wing [ε, λ, ζ, τ?, σ?]; one on the body [γ]; and one, the largest, on the tail [α]. The total is fourteen.

Poetic Astronomy 2.8

The Greeks call this constellation *kyknos* ["swan"], but many, ignorant of its history, give it the name of the genus of birds, *ornis* ["aves"]. The following origin is attributed to it by tradition: Jupiter, when he became enamored of Nemesis and could not persuade her to

lie with him, fulfilled his desire by the following ruse. He ordered
Venus to take on the likeness of an eagle and to follow him; he himself
assumed the form of a swan, and, pretending to flee the eagle, sought
safety with Nemesis, settling himself in her lap. Nemesis did not spurn
him but held him in her arms until she fell asleep. As she slept, Zeus
ravished her then flew away, and, because he appeared to men to fly
high in the sky, it was believed he was placed among the stars. So that
this belief might not be disproved, Jupiter, reflecting what had
occurred, placed in the heavens a swan in flight, pursued by an eagle.
Nemesis, however, as one connected with the race of birds, in due
course brought forth an egg which Mercury carried away to Sparta and
threw into the lap of Leda, as she was sitting. From this egg was born
Helen, who surpassed all other women in beauty, and whom Leda
called her daughter. Others, however, say that Jupiter changed himself
into a swan and lay with Leda. We leave this question unresolved.

This figure has one bright star on the head; another of equal
brightness on the neck; five on each wing; one on the body; one over
the tail. The total of stars is fourteen.

Commentary

The birth of Helen from an egg laid by Nemesis after she consorted
with Zeus is a variant of the more common story according to which
Zeus, in the form of a swan, consorted with Leda, who then bore
Helen, or else produced one or more eggs from which Helen and her
siblings were hatched. A third story, which combines the previous two,
recounts that a shepherd found the egg laid by Nemesis and brought
it to Leda, who watched over it until Helen was hatched, then raised
the girl as her own daughter.[1]

Nemesis, the goddess of divine retribution, appears first in Hesiod,
where she is personified and begins to assume the connotation of
divine indignation and retribution.[2] In later epic, Nemesis is an
anthropomorphized figure who mates with Zeus. In her anthropomorphic form, Nemesis was worshipped at Rhamnus in Attica from at

least the fifth century B.C.E. The cult of Nemesis at Rhamnus super-seded an earlier local cult probably associated with a divinity of the earth-mother type, who appears to have been particularly associated with life and death and, if she bore any resemblance to the goddesses called Nemesis who were worshipped at Smyrna, with vegetation. Farnell argues that the name of the goddess originally worshipped at Rhamnus was Artemis or Artemis Aphrodite, and that the name Nemesis in time became attached to her as a title or epithet. From the worship of this earlier Artemis Nemesis, the later cult of Nemesis developed. While the goddess Nemesis is associated with divine retribution, the meaning of the epithet "Nemesis," when it was attached to the primitive Artemis cult, is not known with certainty, since divine retribution and the moral concept involved in the idea were not normally a function of either Artemis or Aphrodite. Cook, on the other hand, argues that Nemesis was originally a goddess of the woodland to be identified with the Roman Diana Nemorensis and that she "became a goddess of vengeance simply through an illogical but almost inevitable confusion with the abstract substantive *nemesis*, meaning righteous wrath."[3]

The Greeks originally identified the constellation as a bird, without reference to its species. Later, the "bird" was identified with a swan, either because the constellational bird had a long, swan-like neck, or because a swan figured in the first myth to be attached to this constellation. The Babylonians represented the figure in this constel-lation as a four-footed animal.[4]

Two myths in addition to that given by Ps-Eratosthenes and Hyginus are connected with the Swan. According to one, the swan, a musical bird, was changed into a constellation and located near the Lyre as a tribute to Apollo; according to the other myth, the constel-lation represents Cygnus, the cousin of Phaethon and king of Liguria, to whom Apollo granted the gift of song.[5]

The Swan is comprised of fourteen stars according to Ps-Era-tosthenes, Hyginus, and Hipparchus, seventeen according to Ptolemy.

Delphinus

The Constellations 31

This is how the dolphin came to be placed among the stars. When Poseidon wished to make Amphitrite his wife, she, out of modesty, fled to Mount Atlas, eager to preserve her virginity. Most of the Nereids followed her into hiding. Poseidon sent out many searchers to find her, among them the dolphin. While wandering about the islands of Atlas, the dolphin discovered Amphitrite, whereupon it announced her discovery and brought her to Poseidon, who married her and decreed the highest honors of the sea for the dolphin. He declared the dolphin sacred and placed an image of it among the stars. Those who wish to please Poseidon represent him holding the dolphin in his hand, and thus render to the dolphin the highest possible honor for the service it provided. Artemidorus tells the story in his elegies on Eros.

The Dolphin has one star on the mouth [?]; two on the dorsal fin [α, β]; three on the ventral fins [γ, δ, η?]; one on the back [ε]; two on the tail [ι?, κ?]. The total is nine. Because the number of stars [in Delphinus] corresponds to the number of Muses, this animal is said to be fond of music.

Poetic Astronomy 2.17

The reason why the dolphin was placed among the stars is recounted by Eratosthenes and others: when Neptune wished to take Amphitrite as his wife, she preferred to maintain her virginity and hid near Mount Atlas. Many were sent to find her, among them a certain Delphinus, who, wandering about the islands, finally came upon the maiden and persuaded her to marry Neptune; Delphinus himself conducted the marriage ceremony. For this reason Neptune placed the image of a dolphin among the stars. Furthermore, we see that those who make images of Neptune represent him with a dolphin either in his hand or under his foot, because they judge the dolphin to be pleasing to Neptune.

However, Aglaosthenes, who wrote the *Naxica*, says there were some Tyrrhenian shipmasters who were to take the child Liber, along with some companions of his, to Naxos, and there to give him over to his nymph nurses. Both our writers, in books about the origins of the gods, and also many Greek writers, say that Liber was raised by these nymphs. But, to return to our story, the shipmasters, driven by the hope of ransom, sought to divert the ship. Suspecting this, Liber ordered his companions to sing all together. Hearing this unheard of sound, the Tyrrhenians were so delighted with it that they even began to dance about and, in their delight, unwittingly hurled themselves into the sea, where they were changed into dolphins. Liber, because he wished to provide a reminder of them among the human race, placed the image of one of them among the stars.

Others, however, say this is the dolphin that carried Arion the citharode from the Sicilian Sea to Taenarum. Arion was preeminent in his art and earned his living by touring the islands. His young slaves, judging that there was greater advantage in freedom gained by treachery than in peaceful servitude, considered that if they threw their master overboard, they could divide his belongings among them. Arion, when he perceived their plot, requested, not as master from slaves or as an innocent victim from wicked men, but as a parent from his sons, that he be permitted to put on the robe he had often worn in victory, since there was no one other than himself who could mark his misfortune with a lament. When he obtained his request, he took up his cithara, and began to lament his own death. Drawn by the sound, dolphins from all parts of the sea swam up to hear the singing of Arion. Invoking the power of the immortal gods, Arion threw himself upon the dolphins, and one of them took him on its back and carried him to the shore of Taenarum. Because of this, the statue of Arion which was set up there as a memorial has the image of a dolphin affixed to it, and thus the dolphin was depicted among the stars by ancient astronomers. The slaves, who thought they had escaped from servitude, were driven by a storm to Taenarum, where they were apprehended by their master, who inflicted no small punishment upon them.

The figure has one star on the head; two stars above the head on the neck; three stars appear like wings on the belly; one on the back; two on the tail. The total number of stars is nine.

Commentary

The ancient Greeks considered the dolphin the most philanthropic of all creatures and related numerous stories of its service to man.[1] An example is the rescue of Arion by a dolphin, cited here by Hyginus, and elsewhere by many classical authors, which probably gave rise to the widespread boy-on-a-dolphin motif in classical art.[2]

Although Poseidon's association with dolphins, noted by both Ps-Eratosthenes and Hyginus, is widely represented in art, it is attested by few literary allusions.[3] Similarly, recognition of Amphitrite as Poseidon's official consort is sparse in classical literature. The extant literary tradition of this myth begins with *The Constellations*.[4]

Hyginus offers an interesting variation of the motif by anthropomorphizing the philanthropic dolphin. Although he cites Eratosthenes as a source, the Latin author transforms the discoverer of Amphitrite's hiding place into a man, whom he names Delphinus. Further evidence of an anthropomorphic interpretation of the dolphin in connection with this constellation is provided by the scholiast on Aratus, who

refers the epithet Delphinius to the shape assumed by Apollo when he
led the Cretans to Delphi.[5]

The dolphin was not only viewed as the emblem of philanthropy and
service by the Greeks and Romans; a well established tradition,
attested by ancient authors of natural history as well as poets, also
associates the dolphin with love of music. This tradition accounts for
the correspondence noted in *The Constellations* between the number of
stars in the constellation and the nine Muses.[6]

The number of stars comprising Delphinus is nine according to Ps-
Eratosthenes, Hyginus and Hipparchus; ten according to Ptolemy.

Little is known concerning the history of this constellation, and
there is no evidence to suggest a Babylonian connection.[7]

Draco

The Constellations 3

This is the large dragon that lies between the two Bears. It is said to be the dragon that guarded the golden apples and was slain by Heracles. Hera placed the dragon in the land of the Hesperides as guardian of the apples and granted it a position among the stars. Pherecydes says that at the marriage of Zeus and Hera the gods brought many gifts for the bride and that Gaea came bearing the golden apples. Hera marvelled at the apples and ordered that they be planted in the garden of the gods, which was near Mount Atlas. She placed a dragon of extraordinary size to guard against the daughters of Atlas, who were constantly snatching away the apples. The constellation itself is very large. Above it lies the constellation of Heracles, very prominent by its configuration, and placed there by Zeus as a reminder of the struggle.

The figure has three bright stars on the head [β?, γ?, ν?] and twelve stars, set close together along the body, down to the tail [39, ι, ζ, η, θ, δ, ε, α, κ, λ, χ, ψ]. The total is fifteen.

Poetic Astronomy 2.3

This figure stretches its huge body between the two Bears. It is said to be the serpent that guarded the golden apples of the Hesperides, was slain by Hercules, and placed among the stars by Juno because it was in her service that Hercules sought it out. It is believed to be the serpent that watched over the garden of Juno, for Pherecydes says that when the marriage of Jupiter and Juno took place, Terra came bearing golden apples on branches. Juno admired them and asked Terra to plant them in her gardens, which extended to Mount Atlas. Because the daughters of Atlas continually snatched the apples from the trees, Juno was said to have placed the serpent there as a guard. And this is signaled by the fact that above this figure, the likeness of Hercules appears, as Eratosthenes notes. Thus anyone can deduce that this figure is, indeed, to be identified as that particular serpent.

Some, however, say this is the serpent thrown by the giants at Minerva when she fought against them, and that Minerva snatched up the serpent as it lay coiled and flung it to the stars, where it was fixed at the top of the firmament. There it appears to this day with coiled body, as if recently placed among the stars.

Draco has a star on each side of the head; two on the eyes; one on the chin, and ten scattered along the rest of the body. Thus there are fifteen stars in all.

Commentary

The Hesperides were the daughters of Night. Although their name suggests that they were daughters of Hesperus ("evening"), Hesiod says they were conceived without a father. The Hesperides and the dragon of this story were the guardians of the tree bearing golden fruit, which grew in a garden generally believed to lie in the West.[1]

The constellation Draco is always identified with a serpent in classical literature, but there is no general agreement as to which serpent is represented. Beginning with Ps-Eratosthenes, a number of authors, including Hyginus and the Scholiast on Germanicus identify Draco as the guardian of the golden apples of the Hesperides.[2] Many other interpretations are also proposed: Python, the serpent slain by

Apollo at Delphi, or the dragon slain by Cadmus, or, according to Hyginus, the dragon hurled into the heavens by Athena. In still other sources, the constellation is said to represent the snake into which Zeus changed himself when he changed his two nurses into bears.[3]

The concept of the guardian serpent is common to both Greek religion and folklore. Serpents guarded springs as well as structures: a sacred snake, representing the guardian spirit of the city, dwelt in the Erechtheum at Athens; all temples of Asclepius were inhabited by sacred snakes, and in many Greek villages today there is a persistent belief that every house has its guardian snake. The guardian serpents of the golden fleece and the golden apples are a manifestation of the original concept turned folk-motif.[4]

The origin of this constellation is apparently Phoenician. There is no evidence of a Babylonian serpent constellation. Several Egyptian temples are believed to have been oriented to the star γ Dra, but it is not certain whether the Egyptians knew of the constellation Draco.

This constellation is comprised of fifteen stars according to Ps-Eratosthenes, Hyginus, and Hipparchus, thirty-one according to Ptolemy.

Eridanus

The Constellations 37

The River has its source at the left foot of Orion. According to Aratus, it is called Eridanus, although he provides no evidence for this. Others say it is most appropriately identified as the Nile, for that is the only river to have its source in the south. The figure consists of numerous stars. Below it lies the star called Canobus [Canopus,] which touches the steering-oars of the Argo. No star appears lower in the sky, and for this reason it is called Perigeios ["earth-encircler"].

The River has a star at its source [λ = β Ori]; three at the first bend [β, ψ, ω]; three at the second bend [τ¹, τ², τ³]; and seven stars from the third bend to its effluence [υ³, υ⁴, g, f, h, θ, ?], which they say are the mouths of the River Nile. There are thirteen stars in all.

Poetic Astronomy 2.32

Some say this is the Nile, others Ocean. Those who wish to identify it with the Nile point out that this is appropriate because of the great size and usefulness of that river. A further reason given is that there is a certain brightly shining star below it called Canopus, and Canopus was an island washed by the Nile river.

3.31

There are three stars at the first bend; three at the second; seven from the third bend to the last one. The total number of stars is thirteen.

Commentary

Most ancient sources identify this constellation with either the Eridanus or the Nile; a late tradition, however, identifies the constellation as Ocean, or as a part of Ocean.[1]

The Eridanus River was mentioned in Greek literature as early as Hesiod, but its location was seldom agreed upon. Some ancient sources located the Eridanus in northern, some in western Europe; some doubted its existence altogether. It would appear that the name Eridanus was applied to more than one river by the ancients. Greek authors from the time of Pherecydes identified the Eridanus with the Po, although a tradition identifying it with the Rhone persisted. In Latin literature, the Eridanus was invariably identified with the Po River and its numen was personified as the river-god Eridanus. The Eridanus River was connected with amber in both Greek and Latin literature. In ancient times, amber was transported to the Mediterranean from the lands around the Baltic Sea via three main routes which utilized the river systems of northern and central Europe. The central route led over the Brenner Pass into the Po valley.[2]

In Greek myth, the Eridanus was the river into which Phaethon fell when he was struck by the thunderbolt of Zeus. His sisters, the Heliades, mourned for him until they were turned into amber-dropping poplar trees beside the river.[3]

Ps-Eratosthenes and Hyginus identify the river of this constellation with the Nile, presumably because the constellation lies in the southern part of the sky and the Nile lay in the southern part of the Greek world. The Egyptian origin of the name Canobus is attested by the existence of a town, Canopus, which lay twelve miles northeast of Alexandria, on the westernmost (Canopic) mouth of the Nile. Canopus appears in Greek myth as the name of the helmsman of Menelaus, who was supposed to have died of a snake bite at the site of the city of Canopus, which Menelaus founded in his memory.[4]

This constellation is probably of Euphratean origin. In the Euphratean representation of the heavens, there was an area of the southern sky called The Sea, which was fed by two streams of water: the River, and the outflow of the constellations Pisces, Delphinus, Piscis Austrinus, Cetus, and Capricorn. The River appears on numerous Euphratean cylinder seals. In classical and post-classical artistic representations, the River is shown flowing from the left foot of Orion to Cetus, then doubling back to the rudder of the Argo, and from there doubling back again.[5]

The star Canopus was believed by the Greeks to be the lowest visible star and for this reason was called Perigeios. In antiquity, the star was apparently not visible from Greece north of Rhodes. Even from Alexandria, where many of the Greek astronomers made their observations, Canopus would not have appeared very far above the horizon.

Ps-Eratosthenes notes that this constellation-figure consists of "numerous" stars, but, like Hyginus, lists only thirteen. According to Ptolemy, the River was comprised of thirty-four stars.[6]

Galaxy

The Constellations 44

This is one of the heavenly circles, known as the Galaxy. Now, it was not possible for the sons of Zeus to share in heavenly honor before one of them had been nursed by Hera. And so, it is reported, Hermes brought Heracles shortly after his birth and placed him at Hera's breast, and the infant was nursed by her. When Hera discovered the trick, she pushed Heracles away and the remaining milk was spilled, forming the Galaxy.

Poetic Astronomy 2.43

There is a certain circle among the stars, white in color, which some call "milky." Eratosthenes says that on one occasion, Juno unknowingly suckled the infant Mercury, but when she recognized him as Maia's son, cast him away from her, and thus the brightness of the spilled milk appeared among the stars.

Others say that [the infant] Hercules was placed at Juno's breast as she slept and that when she awoke, what we recounted above occurred. Others say that Hercules, in his great eagerness, took so large a quantity of milk that he was not able to contain it in his mouth, and that the [galactic] circle shows what was spilled from his mouth. Others say that at the time when Ops brought to Saturn a stone in place of the child she had borne, he ordered her to offer it milk. When she squeezed her breast, the milk that flowed forth formed a circle, as we recounted above.

Commentary

The galactic circle was one of the eleven heavenly circles distinguished by the ancient Greeks. In addition to the galactic circle, these included the equator, the tropic of Cancer, the tropic of Capricorn, the Arctic Circle, the Antarctic Circle, the horizon, the meridian, the zodiac and the two colures.[1]

The Galaxy was identified by Greek authors as early as Parmenides; however, the few myths associated with its formation occur no earlier than *The Constellations*. In addition to the two myths recounted above, there is one myth that explains the Milky Way as formerly marking the path of the Sun, who changed his course in abhorrence at the deed of Thyestes, and another that explains the Milky Way as the ashes of the scorched heavens left in his path by Phaethon.[2]

Poetic fancy saw in the Milky Way a road, either the road of the gods, or the road beside which stood the palaces of the gods, or the road traveled by the souls of the dead, or the path of the Sun.[3]

The Milky Way was known to the Babylonians, who saw in it a serpent or a rope. The Egyptians saw in the Milky Way a reflection of the River Nile. The designation of the Galaxy as Milk or Milky was, apparently, original with the Greeks.[4]

Gemini

The Constellations 10

This figure is said to be the Dioscuri, who were raised in Laconia, gaining distinction and surpassing all others in their brotherly devotion to one another, for they argued neither about the kingship nor about any other thing. Zeus, wishing to commemorate their comradeship, named them the Twins [Gemini] and placed them in the same place among the stars.

The Twin who rests on the Crab [Cancer] has one bright star on his head [α]; one bright star on each shoulder [ι, ?]; one star on the right elbow [57?]; one on the right hand [74?]; one on each knee [ε, κ Aur?]; and one on each foot [ν, η]. The total is nine. The other Twin has one bright star on the head [β]; one bright star on the left shoulder [κ]; one star on each breast [58, ?]; one on the left elbow [?]; one on the edge of the hand [74?]; one on the left knee [ζ]; one on each foot [?, ξ]; and one, called Propous ["fore-foot"], under the left foot [γ]. The total is ten.

Poetic Astronomy 2.22

Most astronomers say these are Castor and Pollux, who were the most loving of all brothers, because they neither contended for the kingship, nor engaged in any act without mutual counsel. Jupiter is said to have placed them among the most prominent stars on account of their dutifulness. Neptune, too, with similar intent, had rewarded them, for he gave them the horses they rode and the power to save the shipwrecked.

Others say the figures are Hercules and Apollo; some even say Triptolemus, about whom we spoke earlier, and Iasion, who were both beloved of Ceres and placed among the stars. But those who tell of Castor and Pollux also say that Castor was killed in the town of Aphidnae when the Spartans were waging war against the Athenians. Others say that he died in the war waged by Lynceus and Idas against Sparta. Homer says that Pollux granted his brother half of his own life, so that each of them shines on alternate days.

3.21

The one closest to Cancer has one bright star on the head, a bright star on each shoulder; one on the right elbow; one on the same hand; one on each knee; one on each foot; another on the head; one on the left shoulder; one on each breast; one on the left elbow; one on each hand; one on the right knee; one on each foot; and one, called Propous, under the left foot.

Commentary

The Dioscuri, Castor and Polydeuces (Pollux), were either the sons of Zeus and Nemesis or Leda—and, thus, brothers of Helen—or else Polydeuces was the son of Zeus and Castor of a mortal, Tyndareus. When the mortal twin, Castor, was slain in battle, the immortal twin, Polydeuces, conceded one-half of his own immortality to his slain brother, so that they both shared life and death alternately. There is ample evidence that Castor and Polydeuces were Laconian heroes, known as the Tyndaridae, whose cult spread over most of Greece and even to Sicily.[1]

In many societies, twins and their mother are connected with some type of tabu and the twins themselves are often called "children of the sky." Twins may symbolize the dual (day/night) aspect of the sky, whether this symbolism is expressed in a contrast between sun and moon, or, as in the case of the Dioscuri, between evening-star and morning-star. The connection of the Dioscuri with the sky is evident in the earliest Greek literature, while their stellar character, possibly a derivative of that connection, first appears in Euripides.[2]

As Hyginus notes, the Dioscuri were associated with Poseidon and received from him gifts reflecting Poseidon's overlordship of the seas as well as his association with horses. The Dioscuri were represented on coins of Tarentum in Magna Graecia, c. 315 B.C.E., as well as on coinage of the Hellenistic kingdoms of the Eastern Mediterranean. Beginning about 269 B.C.E., the Dioscuri appeared regularly on coinage of the Roman Republic. On both Roman and Hellenistic coinage, the Dioscuri were depicted as mounted horsemen, wearing caps upon which a star rests. The Dioscuri were believed to come to the assistance of mariners in distress, a function later assumed by the Christian Saint Nicholas. In antiquity, the atmospheric phenomenon now known as St. Elmo's fire was associated with the Dioscuri and was interpreted as a favorable omen when it appeared with two flames, but as unfavorable if it appeared with only one flame.[3]

The constellation Gemini is always associated with two figures in Greek literature. The most common identification is with Castor and Polydeuces, but Amphion and Zethus are also mentioned, along with Heracles and Apollo, Heracles and Theseus, Triptolemus and Iasion, the Great Gods of Samothrace, and Phaon and Satyrus. The Greek constellation is probably borrowed from the Euphratean constellation of the Great Twins.[4]

The number of stars in Gemini is nineteen according to Ps-Eratosthenes, Hyginus, and Hipparchus, eighteen according to Ptolemy.

Hercules

The Constellations 4

The figure standing on the Dragon is said to be Heracles. He is clearly standing, wrapped in the lion's skin, his club raised. It is recounted that when Heracles went to fetch the golden apples, he slew the dragon that was guarding them. The dragon had been placed there by Hera in order to oppose Heracles. The feat was accomplished in the face of very great danger, and Zeus, judging it worthy of memory, placed an image of it among the stars. The serpent's head is raised high; Heracles is astride the serpent and holds it pinned with one knee while he steps on its head with the other foot. His right hand, which holds the club, is extended as if he were about to strike; he wears the lion's skin over his left arm.

The figure has one bright star on the head [α]; one bright star at the right bicep [γ]; one bright star on each shoulder [β, δ]; one on the left elbow [μ]; one at the edge of the hand [o?]; one on either side below the ribs [ζ, ε], with the brighter one on the left; two on the right thigh [η, σ]; one at the knee [τ]; two on the leg [υ, χ]; one on the foot [74]; one, which is called the Club, over the right hand [ω]; four on the lion's skin [o?, ν?, ξ?, λ?]. The total is nineteen.

Poetic Astronomy 2.6

Eratosthenes says the figure located above Draco is Hercules, about whom we spoke earlier. He is prepared as if for a struggle, holding the lion's skin in his left hand and the club in his right hand. His struggle is to slay the serpent that guarded the Hesperides and was believed never to close its eyes in sleep—all the more proof that it was placed there as guard. The story is also recounted by Panyasis in the *Heraclea*. Jupiter, admiring the struggle, represented it among the stars. Draco's head is erect; Hercules, on his right knee, attempts to step on the right side of Draco's head with his left foot; his right arm is extended as if to strike, his left is outstretched holding the lion's skin, so that he appears to be struggling mightily. Although Aratus says that no one can prove who this figure is, we will try to say something approaching the truth.

As we said earlier, Araethus says that this is Ceteus, the son of Lycaon and father of Megisto, and that he appears to be lamenting the changing of his daughter into a bear. He is on his knees and holds out both hands to the heavens in prayer that his daughter might be restored to him. Hegesianax, however, says this is Theseus, who appears to be raising the stone at Troezen, for Aegeus was believed to have placed the Ellopian sword under the stone and instructed Aethra, Theseus's mother, not to send Theseus to Athens before he was strong enough to raise the stone and bring the sword to his father. Thus he appears struggling to raise the stone as high as he can. For this reason some say the Lyre, which is the closest constellation, is the lyre of Theseus, since he, being accomplished in all the arts, had also studied the lyre. Anacreon, too, mentions this: "Near Theseus, the son of Aegeus, is the Lyre."

Others, however, say it is Thamyris, blinded by the Muses and supplicating them on his knees; others that it is Orpheus, slain by the Thracian women because he had looked upon the rites of Liber. In his tragedy titled *Prometheus Unbound*, Aeschylus says this figure is Hercules fighting not with Draco, but with the Ligurians. For he says that when Hercules drove away the cattle of Geryon, his journey brought him through the Ligurian territory. The Ligurians banded together in an effort to take away his cattle and he shot many of them with arrows. When his arrows were exhausted, Hercules, overwhelmed by the number of the barbarians and by his own lack of arms, knelt, already heavily wounded, and Jupiter in pity came to the rescue of his son, placing a large quantity of stones around him, with which Hercules defended himself and chased away the enemy. Jupiter then placed an image of the fighter among the stars.

Some say this is Ixion, his arms bound because he tried to overpower Juno; others say it is Prometheus bound on the Caucasus Mountains.

The constellation has one star on the head; one on the right arm; a bright star on each shoulder; one on the left elbow; one on each side, the one on the left being brighter; two on the right thigh; one on the knee; one on the back of the knee; two on the leg; one on the foot; one above the right hand, which is called the Club; four on the left hand, which some say represent the lion's skin. There are nineteen stars in all.

Commentary

This constellation—also called the Kneeler (*Engonasin*) by the Greeks—was usually connected with the constellation Draco in mythology. The Kneeler is Heracles, according to most sources, and Draco is the serpent he slew to gain the golden apples of the Hesperides in one of the last labors imposed on him by King Eurystheus.[1] The Greeks were not the first to identify Draco and the Kneeler with a serpent and its slayer. Phoenician tradition associated these two constellations with a dragon and the Sun God who slew the dragon.

This constellation figure was not always identified with Heracles by the Greeks. Aratus refers to the constellation simply as the Kneeler; subsequent Greek and Roman authors, including Hyginus, identify the figure with Theseus, Thamyris, Orpheus, Ixion, Prometheus, Atlas, Tantalus, or Ceteus.[2]

The number of stars in this constellation is nineteen according to

Ps-Eratosthenes and Hyginus, twenty-four according to Hipparchus, and twenty-nine according to Ptolemy.

Hydra,
Crater, Corvus

The Constellations 41

This constellation is well-known from a famous episode. Each god has a bird as an attribute, and the crow is the attribute of Apollo. Once, when the gods were preparing a sacrifice, the crow was sent to bring the libation from a certain spring which was considered most sacred before wine was invented. Seeing a fig tree with unripe fruit near the spring, the crow waited until the figs were ripe. After a number of days, the crow ate the ripe figs, then realized his misdeed. Snatching up the water-snake from the spring, the crow brought it back, along with the water-cup, alleging that the water-snake had daily been consuming the water in the spring. Apollo, however, knowing the truth, imposed on the crow the punishment of thirsting among men for a long period of time, as Aristotle explains in his treatise on animals; Archelaus says the same thing in his *On Peculiar Things*. In order to provide a clear warning about sinning against the gods, Apollo placed among the stars the image of the Water-Snake [Hydra], the Water-Cup [Crater] and the Crow [Corvus], and depicted the latter as if prevented from drinking or approaching [the Water-Cup].

The Water-Snake has three bright stars at the top of its head [σ, δ?, ε?]; six on the first coil [θ, ι, τ¹, τ², ω, α], of which the last is bright; three on the second coil [κ, υ¹, υ²]; four on the third [μ, ν, φ, ?]; two on the fourth [β Crt, χ¹]; nine faint stars from the fifth coil to the tail [ξ, o, β, γ, π, ?, ?, ?, ?]. The total is twenty-seven.

On the tail of the Water-Snake is the Crow, facing west. The Crow has one faint star on the beak [α Crv]; two on the wings [γ Crv, δ Crv]; two on the tail [η Crv?, ?]; one on the edge of each foot [β Crv, ?]. There are seven stars in all.

The Water-Cup lies at some distance from the Crow, and is tipped toward the knees of the Maiden [Virgo]. The Water-Cup has two stars on the rim [ε Crt, ζ Crt]; two faint ones on each handle [η Crt, θ Crt, ?, ?]; two on the middle [γ Crt, δ Crt]; two at the base [α Crt, ?]. The total is ten.

Poetic Astronomy 2.40

Concerning this constellation, tradition has provided the following origin. The crow, enjoying Apollo's protection, was sent to a fountain to fetch pure water for Apollo's sacrifice when he saw some trees with their figs not yet ripe and perched in one of the trees, waiting for the figs to ripen. After a few days, when the figs were ripe and the crow had eaten his fill, Apollo, awaiting the crow, saw him struggling in hasty flight with a full water-cup. Because of the offense of delay, Apollo, who was forced to use other water while the crow tarried, inflicted the following punishment. While figs are ripening, crows cannot drink, because during those days their throat is affected. Thus, because he wanted to signal the thirst of the crow, Apollo placed among the stars the Water-Cup and below it Hydra, to delay the thirsting Crow. The Crow appears to be shaking Hydra's tail with his beak, in order to gain access to the Water-Cup.

Istrus, however, and many others say that this is Coronis, the daughter of Phlegyas, and that she was the mother of Asclepius by Apollo; later, Ischys, the son of Elatus, lay with her. When the crow saw this, he told Apollo, and Apollo, in return for the unwelcome message, slew Ischys with his arrows, and changed the crow's color, which was formerly white, to black.

Concerning the Crater, Phylarchus tells this story. In the Chersonesus, which is located near Troy, where the tomb of Protesilaus lies, there is a city called Eleusa. During the reign of a certain Demiphon, widespread devastation and an unexpected plague befell the city. Demiphon, greatly perturbed, sent to the oracle of Apollo to inquire how the devastation might be halted. The response of the oracle was that a maiden of noble birth must be sacrificed each year on the altar of the city's gods. Demiphon, choosing the maidens by lottery, sacrificed all other daughters save his own, until one of the well-born citizens complained of the practice of Demiphon. This man said he would not allow his daughter to be part of the lottery unless the daughters of the king were part of it as well. The king was angered and, selecting that man's daughter without a lottery, put her to death. The maiden's father, Mastusius by name, pretended at the time that he would not be angry since the deed was done on behalf of their country, for the lot might have fallen to her later, and she might have perished nonetheless. After a few days, the father of the maiden lulled the king into forgetfulness, then, when he had shown himself to be most kindly disposed toward the king, claimed that he was preparing a solemn sacrifice and invited the king and his daughters. The king, not suspecting that anything untoward was about to happen, sent his daughters ahead, as he was occupied with matters of state and planned to come later. When what Mastusius had greatly hoped for happened, he slew the king's daughters and, mixing their blood with the wine in the wine-jar, ordered that it be offered to the king to drink as he approached. When the king looked for his daughters and discovered what had happened to them, he ordered that Mastusius be thrown into the sea, along with the wine-jar. For that reason the sea into which he was thrown was called Mastusian in his memory, and the port is to this day called Crater ["wine-jar"]. The ancient astronomers configured it among the stars so that men might be reminded that no one can profit from an evil deed, and that evil deeds cannot be forgotten.

Some, agreeing with Eratosthenes, say this is the wine-jar that Icarius used when he was showing men the uses of wine; others say it is the vessel in which Mars was imprisoned by Otus and Ephialtes.

3.39

Hydra has three stars on the head; six stars on the first coil after the head, the last one of which is bright; three on the second coil; four on the third; two on the fourth; and nine stars from the fifth coil to the tail, all of them faint. The total is twenty-seven.

Corvus has one star on the throat; two on the wing; two beneath the wing toward the tail; one on each foot. The total is seven.

Above the first coil after the head is the Crater, which has two stars on its rim; two faint stars under the handles; two in the middle; two on the base. The total is ten.

Commentary

Ps-Eratosthenes implies that both this constellation and the myth which he relates about it are well-known. His latter contention, at least, is not supported by what has survived of Greek literature. Although the crow is mentioned by Greek authors as the messenger of Apollo and figures in the myth of Coronis, there is no Greek authority earlier than *The Constellations* which refers to the myth of the crow and the figs, nor does this myth occur outside an astronomical context in the few later sources which relate it.[1]

As in the case of the Perseus-Andromeda group, the constellations of the Hydra-Crater-Corvus group were placed among the stars to commemorate an offense against the gods rather than a noble deed. Indeed, the story of the crow and the figs is as much an aetion explaining the crow's thirst in summer as it is a constellation myth. This same story, with one minor variation, is offered by Aelian as the reason for the crow's thirst in summer.[2]

Between them, Ps-Eratosthenes and Hyginus allude to most of the myths associated with this constellation-group by classical authors.[3]

Although it may be possible to trace the Crow and the Hydra to Euphratean constellations, there is little evidence for associating the two figures, as Ps-Eratosthenes and Hyginus have done. The Crow may have its origin in the demon-ravens of Euphratean mythology or in the constellation of the Storm-Bird.[4]

The Hydra is comprised of twenty-seven stars according to Ps-Eratosthenes, Hyginus, Hipparchus, and Ptolemy. The Crater consists of ten stars according to Ps-Eratosthenes, Hyginus, and Hipparchus, nine according to Ptolemy. All three authorities give the number of stars in the Crow as seven.

Leo

The Constellations 12

The Lion [Leo] is one of the bright constellations. It is believed this sign was honored by Zeus because the lion is the king of beasts. Some say the constellation bears witness to the first labor of Heracles, who, seeking fame, killed this animal without a weapon, strangling it with his hands. Pisander of Rhodes tells the story. Heracles thereafter wore the lion's skin, as having performed a notable feat. This is the lion he slew at Nemea.

The Lion has three stars on its head [ϵ, μ, λ?]; one on the chest [α]; two below the chest [31, ν]; one bright star on the right foot [ξ]; one on the middle of the belly [46]; one under the belly [53?]; one on the haunch [δ]; one on the back knee [ψ]; one bright star on the edge of the foot [υ?]; two on the neck [γ, ζ]; three on the back [41, 54, η?]; one in the middle of the tail [?]; one bright star at the end of the tail [β]; and one on the belly [52?]. The total is nineteen. Above the lion's tail seven faint stars are visible in the shape of a triangle [15, 7, 23, ?, ?, ?, ?]. These are called the Lock of Berenice Euergetis [Coma Berenices].

Poetic Astronomy 2.24

This constellation is said to have been created by Jupiter, because the lion is considered the king of all the wild animals. Others say in addition that this was the first labor of Hercules and that he killed the lion without weapons. Pisander and many others have written about this.

Above this figure, near Virgo, is a group of seven stars configured in a triangle on the tail of Leo; these are said by the mathematician Conon of Samos and by Callimachus to be the Lock of Berenice. When Ptolemy married his sister Berenice, daughter of Ptolemy and Arsinoe, and after a few days went off to war in Asia, Berenice vowed that if Ptolemy returned victorious, she would make a votive offering of a lock of her hair. After his return, she placed the votive lock of hair in the temple of Venus Arsinoe Zephyritis. On the following day, the lock could not be found and the king was greatly agitated. Conon the mathematician, wishing to gain the favor of the king, proclaimed that the lock had been placed among the stars, and he pointed out seven stars that were not part of another constellation, representing these as the lock.

Some, including Callimachus, say that this Berenice raised horses and used to race them at Olympia. Others say further that, once, Ptolemy, Berenice's father, was frightened by the number of the enemy and fled, but his daughter, being accustomed to riding, leapt on a horse and marshaled the remaining soldiers, killing many of the enemy, and putting the rest to flight. On account of this, Callimachus called her "great-souled."

Eratosthenes says Berenice ordered that the dowry left by their parents to the maidens of Lesbos be returned to them, since no one had released it, and she instituted among them the legal process of recovery.

3.23

The figure has three stars on the head; two on the neck; one on the chest; three between the shoulders; one in the middle of the tail;

another on the end; two under the chest; one bright star on the front foot; one bright star on the belly, and another large one below it; one on the loin; one on the back knee; one bright star on the back foot. The total number of stars is nineteen.

Commentary

The slaying of the Nemean lion by Heracles was a popular subject both in art and in literature beginning in the sixth century B.C.E. According to tradition, the Nemean lion was the offspring of earth-born monsters, or else fell from the sky or the moon and was reared by Hera. Both the artistic and literary themes of a man struggling with a lion appear to be of Near Eastern origin. Whether the tradition that the Nemean lion fell from the sky or the moon points to an astronomical origin for the theme itself is an interesting question. Although lions are mentioned often by Greek authors, evidence for the physical presence of lions in Greece is slim.[1]

Both Ps-Eratosthenes and Hyginus identify the figure in this constellation as the Nemean lion; however, both authors also report another tradition, which did not identify the constellation with a particular lion, but with the lion in general, as the king of beasts.[2]

The lion constellation appears frequently on cylinder seals and monuments from the Euphratean area in the form of a lion struggling with a hero (usually Gilgamesh) or with the Sun God. The association of the lion with the sun was apparently due to the fact that the lion was the eighth of the Babylonian zodiacal signs and coincided with the summer solstice.[3]

The number of stars in Leo is nineteen according to Ps-Eratosthenes and Hipparchus, twenty-seven according to Ptolemy. The star α Leonis (Regulus) was called *Basiliskos* ("small king") in Greek, and "the star of the King" by the Babylonians.

Coma Berenices was named by the astronomer Conon *c.* 245 B.C.E. in honor of Queen Berenice II Euergetes (273–221 B.C.E.), after the return of her husband, Ptolemy III Euergetes, from a successful campaign in Syria. Hyginus's mention of her horsemanship and her establishing of the legal process of recovery are not attested elsewhere.[4]

The star-cluster comprising Coma Berenices is variously described in Greek literature both before and after Conon. Ps-Eratosthenes refers to these stars as both the Lock of Berenice and the Lock of Ariadne. According to other authors, the stars of Coma Berenices were thought to be the mane of the Lion, an ivy-leaf, a grape cluster, or a spindle. Aratus and his translators knew of no name for these stars.[5]

Lepus

The Constellations 34

This is the hare that is part of the so-called "hunt." Hermes is said to have placed this animal among the stars on account of its speed. It is believed to be the only four-footed animal that conceives multiply, giving birth to some offspring, while it bears others in the womb; this is explained by the philosopher Aristotle in his treatise on animals.

The Hare has one star on each ear [ι, ν]; three on the body [α, β, ζ], of which the one on the back is bright; and one on each back leg [δ, γ]. The total is seven.

Poetic Astronomy 2.33

The Hare is said to be fleeing the dog of the hunter Orion. For since Orion was, appropriately, depicted as a hunter, it was desirable to show what he was hunting, and a fleeing hare was placed at his feet. Some say that the Hare was placed among the stars by Mercury and that hares have the exceptional power among quadrupeds of giving birth and being pregnant at the same time.

Those who disagree with this view say that it was inappropriate to represent so noble and great a hunter, about whom we spoke earlier in connection with the sign of the Scorpion [2.26], as hunting a hare.

They also reproach Callimachus because, in praising Diana, he said that she delighted in the blood of hares and hunted them; and so, they represent Orion as struggling with the Bull.

The following story about the Hare is handed down by tradition. In ancient times, there were no hares on the island of Leros. A certain young boy of that city, developing an interest in that animal, brought a pregnant female hare from abroad and took great care of it until it gave birth. When it had given birth, many townspeople became interested and took to raising hares, acquiring them partly through purchase and partly as gifts. Within a short time, there was such a number of hares born that the entire island was said to be overrun with them. When their masters gave them nothing to eat, the hares attacked the crops and consumed everything. The inhabitants, facing calamity since they were suffering from famine, took counsel together and eventually drove the hares from the island. And so, at a later time, the image of a hare was placed among the stars to remind men that nothing in life is so desirable that it does not entail greater pain than pleasure in its aftermath.

3.32

The Hare has one star on each ear; two on the body; one on the back; one on each front foot. The total number is seven.

Commentary

Ps-Eratosthenes and Hyginus constitute the principal references to this constellation in classical literature. These authors are also unique in alluding to the "hunt" of which the Hare is a part, together with Orion and Procyon. The overrunning of the island of Leros by hares, recounted by Hyginus, is a curious story, more morality tale than aetion: no Greek god is an agent in the story, and there is no named protagonist. A plague of hares on the island of Astypalaea (an island in the vicinity of Leros) in the reign of Antigonus Gonatas (258–239 B.C.E.) is mentioned by Athenaeus, and may provide a historical basis for the tale.[1]

According to both Ps-Eratosthenes and Hyginus, the Hare was placed among the stars by Hermes on account of its swiftness. The connection of Hermes with hares is remote, at best: the hare is not among the animals normally associated with Hermes. The messenger of the gods was, however, known for his swiftness of foot—two of his epithets were *okypedilos* ("swift-footed") and *okys* ("swift")—and hence might be thought to have some connection with animals possessed of the same power.[2]

As noted by *The Constellations*, Aristotle makes several references to superfetation in hares.[3]

Lepus would appear to be of Greek origin, as no hare-constellation has been identified in Babylonian records. There is, however, ample evidence, both in Babylonian and numerous other sources, which represents the hare as a lunar symbol.[4]

The number of stars in Lepus is seven according to Ps-Eratosthenes and Hyginus, eighteen according to Hipparchus, and twelve according to Ptolemy.

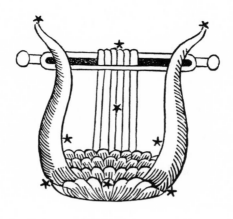

Lyra

The Constellations 24

This is the ninth in order of the constellations and is the lyre of the Muses. It was invented by Hermes from the tortoise shell and the cattle of Apollo. The lyre had seven strings, either from the seven planets or from the seven daughters of Atlas. In his turn, Apollo received the lyre, adapted song to it and passed it on to Orpheus, who was the son of the Muse Calliope. Orpheus increased the number of strings to nine, corresponding to the number of the Muses. He was held in ever greater esteem among men so that the story spread about him that he charmed trees, rocks, and wild animals with his song. When he descended to the Underworld on account of his wife and observed the state of affairs there, Orpheus ceased to honor Dionysus, through whom he had gained glory. Instead, he believed Helius ["Sun"] to be the greatest of the gods, calling him Apollo. Orpheus would arise at night just before the dawn and climb Mount Pangaeum to await the sunrise, so that he might before all else look upon Helius. Dionysus became so angered that he sent the Bassarides against him, as Aeschylus the tragedian recounts. These tore Orpheus limb from limb and scattered his members in different places. The Muses collected the limbs and buried them at the place called Leibethroe.

Having no one to whom they could give the lyre, the Muses asked Zeus to change it into a constellation, so that there might be a memorial to Orpheus and to themselves among the stars. Zeus granted their wish and the lyre was thus placed in the heavens. It bears the sign of Orpheus's misfortune by setting in each season.

The Lyre has one star on each horn [ε, ζ]; one at each end of the bridge [β, γ]; one on each arm [η?, θ?]; one on the crossbar [δ]; and one bright white star at the base [α]. The total is eight.

Poetic Astronomy 2.7

According to Eratosthenes, the Lyre was placed among the stars for the following reason. First fashioned by Mercury out of a tortoise shell, the lyre was handed down to Orpheus, son of Calliope and Oeagrus, who was diligent in his application to it. It was believed that by his art, Orpheus was able even to attract wild beasts to hear his music. Grieving over the death of his wife Eurydice, Orpheus went down to the Underworld and there praised all the gods' offspring with his song, except for Liber, passing him by in forgetfulness, just as Oeneus overlooked Diana in his sacrifice. Thus, later on, while Orpheus was delighting in song on Mount Olympus, which divides Macedonia from Thrace, or, as Eratosthenes says, on Mount Pangaeum, Liber set the Bacchae upon him and they tore Orpheus limb from limb. Others say that this befell him because he had spied on the sacred rites of Liber. The Muses collected his members and buried them, and, according the highest honor they could, placed the figure of a lyre among the stars in his memory, with the consent of Apollo and Jupiter, for Orpheus had greatly honored Apollo. Jupiter conceded this honor to his daughter [Calliope].

Others say that Mercury, when he invented the lyre on Mount Cyllene in Arcadia, gave it seven strings, from the number of Atlas's daughters, of whom his mother, Maia, was one. Later on, Mercury stole the cattle of Apollo but was captured by the latter. In order to gain Apollo's pardon more readily, Mercury agreed to Apollo's request that the invention of the lyre be credited to him [Apollo], and accepted in return a certain rod. Mercury set off through Arcadia holding the rod in his hand when he saw two serpents, their bodies intertwined, attacking each other. In order to separate them, he placed the wand between them, and so they moved apart. Because of this, he established the wand as an agent of peace.

Many, when they represent the caduceus, show a wand with two serpents winding around it, because Mercury was the initiator of peace. From his example, the wand is used both in athletic and in other types of contests. But to return to our original topic, Apollo accepted the lyre and was said to have instructed Orpheus. Later, when he himself invented the cithara, he gave the lyre to Orpheus.

Many say that when Venus and Proserpina came to seek Jupiter's judgement as to which of them should have Adonis, Jupiter ordained that the Muse Calliope, Orpheus's mother, should decide between them. Calliope ruled that each goddess should possess Adonis for half the year. Venus, angered that Adonis was not given to her alone, caused all the women of Thrace to become enamored of Orpheus and to fight over him so that they tore him limb from limb. His head was carried down to the sea and washed ashore by the waves on the island of Lesbos, whose inhabitants took it and accorded it burial rites. On account of this act, they are considered to be extremely gifted in the musical arts. Orpheus's lyre, as we mentioned earlier, was placed among the stars by the Muses. Some say that because Orpheus introduced the love of young boys, he was perceived by women to disdain them and, for this reason, they killed him.

3.6

The figure has one star on each side of the tortoise shell; one on the tip of each horn (the horns are attached to the shell-like arms); one on the middle of each horn, which Eratosthenes represents as shoulders; one on the shoulder of the shell; one at the bottom of the lyre which appears to be the base for the whole figure.

Commentary

The lyre was believed by the Greeks to be the invention of Hermes. Although the original number of strings is a matter of debate, all sources agree as to the form of the instrument. The lyre consisted of two horns attached to a tortoise-shell base; a transverse piece of wood connected the two horns near the top and served also to fasten the strings, which were plucked either with the fingers or with a plectrum. According to both Ps-Eratosthenes and Hyginus, the lyre originally had seven strings. It is more likely, however, that the lyre originally had fewer than seven strings—either three or four—judging from the information related by several ancient sources, that Terpander (*fl. c.* 650 B.C.E.) added three strings to the original four, thereby forming the heptachord. Archaeological evidence lends further support to that tradition. Vases of the eighth century B.C.E. show five-stringed lyres, while seven-stringed lyres appear for the first time in vases of the following century. Pindar (fifth century B.C.E.) was familiar with an eight-stringed lyre, and Timotheus of Miletus was reputed to have increased the number of strings to eleven in the fourth century B.C.E. The name lyre (*lyra*) seldom occurs in the early Greek writers, who more often use the names cithara (*kitharis*) and phorminx (*phorminx*), leading modern scholars to conclude that, originally, lyra and kitharis were two names applied to the same instrument. By the time of Pindar, enough innovations had been introduced to differentiate the lyre from the cithara, although the two were often confused. The invention of the cithara was attributed by the Greeks to Apollo, who was also closely connected with the lyre in consequence of a bargain struck with Hermes, as related by Hyginus. The lyre was the earliest, and remained one of the principal instruments of the Greeks. It was used mostly as an accompaniment to song.[1]

As indicated in both *The Constellations* and Hyginus, Orpheus was closely connected with music in the minds of the ancients. Music was inseparable from almost every aspect of his legend. Like most of the great musicians of Greek myth, Orpheus was Thracian. He was believed to be the son of Calliope, the Muse of epic poetry. Whether or not the innovation of the nine-stringed lyre can be attributed to him, as suggested in *The Constellations*, Orpheus was named by several ancient sources as the first musician, and subsequent Greek musicians of repute were said to have been his pupils. According to tradition, Orpheus's skill as a musician was so great that his music was believed to exert a magical power: trees, stones and wild animals, as well as men, succumbed to its beauty. Several other skills which can be considered extensions of his musical talents were attributed to Orpheus in antiquity. The Greek term music (*mousike*) included not only song and instrumental music, but also poetry and prophecy. Several ancient authors named Orpheus as the inventor of the hexameter, and the great mass of "Orphic" literature attested to Orpheus's reputation as a poet. Orpheus's prophetic powers were mentioned by several ancient sources, and further attested to by his connection with Apollo.[2]

However great Orpheus's reputation as a musician, poet, and prophet, it was surpassed by his importance as a religious figure. His role in the history of Greek religion and his relationship to Apollo and Dionysus are complicated. Among modern scholars, Orpheus is generally held to be a mortal priestly figure of either the Dionysiac or the Apolline religion, and his role to be that of a religious reformer. Harrison maintains that Orpheus was a man who modified the rites of Dionysus by adding a spiritual dimension to Dionysiac religion. Farnell believes that Orpheus is to be considered a Dionysiac priest-figure who suffered the same fate as his god, and not a divinity in his own right. Guthrie considers Orpheus to be a Greek priest of the Apolline religion who somehow embraced the established Dionysiac cult of Thrace. The relation of Orpheus to Apollo is recognized by all three of the above scholars; however, Harrison de-emphasizes the similarities of the two figures and considers the differences between them to be of greater importance. She argues that any similarities are overshadowed by the air of mysticism surrounding Orpheus, which is completely lacking in the "rationality and lucidity" of Apollo. That the religious teachings of Orpheus were identical with those of the

Dionysiac religion is attested to by Herodotus.[3]

Orpheus's worship of Apollo and identification of the latter with Helius is reported by *The Constellations* to have offended Dionysus and led to Orpheus's death. Ancient literary and archaeological evidence suggests that the identification of Apollo with Helius was not common until Hellenistic times at the earliest. Thus, it would appear that if the story of Orpheus's death at the hands of the Thracian women was current before the Hellenistic age, and the reference to Aeschylus in *The Constellations* shows that it was, then the explanation provided by *The Constellations* for the enmity between Dionysus and Orpheus is only one of several such explanations.[4]

Both Ps-Eratosthenes and Hyginus follow ancient tradition in recounting the descent of Orpheus to the underworld in order to recover his wife, Eurydice. Some scholars maintain that Orpheus was originally an underworld spirit to whom the later story of the descent in search of a lost wife was attached, possibly after his adoption by the Dionysiac cult, since a descent to the underworld would be incongruous with the cult of Apollo. Once he was adopted by the Dionysiac cult, Orpheus began to teach a more or less sacramental religion focused on the afterlife. A cult which promised happiness after death could, appropriately, have a founder with some experience of the underworld. In some artistic representations of Orpheus in the underworld, he is shown not with Eurydice, but alone. The earliest evidence for the descent is a painting of Polygnotus [fifth century B.C.E.], described by Pausanias, while the earliest classical literary source for the Orpheus-Eurydice story may be Vergil. Eurydice herself was probably a Thracian nymph. Some few sources indicate that Orpheus was successful in restoring Eurydice to life, but the traditional account speaks of failure because he broke the tabu against looking back.[5]

After the second loss of his wife, Orpheus was said to have passed a long period in mourning, shunning the company of women. Either for this reason, or because he would not admit women to his rites, or because he no longer worshiped Dionysus, Orpheus incurred the wrath of the Thracian women, who tore him limb from limb. Whatever the precise reasons for Orpheus's death, it appears to have been connected with the cult of Dionysus. There is evidence of a cult of Orpheus at Leibethroe, the traditional site of his death.[6]

The origin of this constellation is obscure. It may have originated in either the Euphratean bird-constellation or the Phoenician lyre-constellation. Some scholars have even suggested an Egyptian origin. The only identification of this constellation by classical authors is with the lyre of Hermes.[7]

The number of stars comprising the Lyre is eight according to Ps-Eratosthenes and Hyginus, ten according to Hipparchus and Ptolemy.

Ophiuchus

The Constellations 6

This is the figure standing close to the Scorpion [Scorpio], and holding a snake with both hands. It is said to be Asclepius, whom Zeus placed among the stars to oblige Apollo. For Asclepius had used his healing art to raise the dead—<most recently, Hippolytus, the son of Theseus. The gods were annoyed that the honors paid them might diminish because of the feat of Asclepius.> It is said that Zeus grew angry and struck Asclepius's house with lightning but later placed him among the stars because of Apollo. <The constellation covers a large area, being above the brightest of all the constellations, the Scorpion, and is easily distinguished.>

Ophiuchus ["serpent-holder"] has one bright star on the head [α]; one bright star on each shoulder [β, ι]; three stars on the left hand [δ, ε, λ?] and four on the right hand [μ, ν, τ, ?]; one star on each hip [?, ?]; one on each knee [η, ζ]; one on the right leg [ξ]; one on each foot [θ?, ρ], with the one on the right foot brighter. The total is seventeen. At the top of the Serpent's head there are two stars [β Ser, γ Ser] [. . .].

Poetic Astronomy **2.14**

Ophiuchus, who is called Anguitenens ["serpent-holder"] by our writers, is located above Scorpio; in his hands he holds a serpent which winds around his body. Many say this is the king of the Getae who live in Thrace, Carnabon by name, who came to power at the time when grain was first introduced to mankind. Now Ceres, when she bestowed her gifts on men, ordered Triptolemus (who was said to have been the first to use the wheel so as not to be delayed in his course), whose nurse she had been, to mount the dragon chariot and to travel about distributing grain to the fields of all nations, so that they and their posterity might more easily be distinguished from the wild beasts by their diet.

When Triptolemus came to the aforementioned king of the Getae, he was at first greeted as a guest, but later, as a most cruel enemy—not as a harmless bestower of benefits—such that he was captured by treachery, and he who was prepared to prolong the lives of others almost lost his own life. For Carnabon ordered one of the dragons to be killed, so that Triptolemus might not hope to find escape in his chariot when he discovered that a plot was in the making. But Ceres came and took the chariot away, hitched another dragon to it, and gave it back to the boy. The king was greatly punished for his attempted misdeed. Hegesianax says that Ceres, as a reminder to mankind, placed Carnabon among the stars with a dragon in his hands that he appears to be strangling. Carnabon lived so wretchedly that he incurred a most welcome death.

Others argue that this is Hercules near the Sagaris river in Lydia, killing a serpent that had caused many deaths and laid waste the region. Hercules was rewarded for this feat by Omphale, who reigned there, and he returned to Argos laden with many gifts. Jupiter placed him among the stars on account of his bravery.

Some say this is Triopas, king of the Thessalians, who destroyed an ancient temple of Ceres when he was constructing a roof for his own house. For that reason Ceres brought upon him a hunger which could never be satisfied with any food. Later, toward the end of his life, he suffered great ills when a serpent was set upon him by Ceres, and earning death, he was placed among the stars by wish of Ceres. And so to this day the serpent, strangling him, appears to be inflicting the punishment he deserved.

But Polyzelus of Rhodes points out that this is Phorbas, who proved to be of great assistance to the Rhodians. When the island of Rhodes was overrun with serpents, so that its inhabitants called it *ophioussa* ["teeming with snakes"], in that horde of beasts there was reportedly one serpent of enormous size, which killed many men. When the ravaged country began to be depopulated, it was said that Phorbas, the son of Triopas and Myrmidon's daughter Hiscilla, was brought there by a storm and killed all the beasts including the large serpent. Because he was very dear to Apollo, Phorbas was portrayed among the stars slaying the serpent, as a reward and a memorial. And so the Rhodians, whenever their fleet comes away from their shore, sacrifice first of all to the coming of Phorbas, in the hope that such an event of unexpected courage should befall their people, as the one that brought Phorbas, heedless of future praise and glory, to the stars.

Many astronomers, however, think this is Asclepius, whom Jupiter placed among the stars for Apollo's sake. For when Asclepius was on earth, he excelled in the medical arts to such a degree that he was not content with alleviating human pain, but aspired to restore the dead to life. And he was said, most recently, to have resuscitated Hippolytus, who was killed because of the inequity of his stepmother and the ignorance of his father, as Eratosthenes relates. Some say that Glaucus, the son of Minos, was brought back to life by Asclepius's skill. On account of this transgression, Jupiter destroyed Asclepius's house with a thunderbolt, but placed him, because of his skill and for the sake of his father Apollo, among the stars, holding a serpent.

Some say that Asclepius holds a snake for the following reason. When he was ordered to heal Glaucus and was imprisoned in a secret place, Asclepius sat considering what to do, staff in hand, and a snake crept up his staff. Asclepius, alarmed, killed the snake with many strokes of his staff as it attempted to flee. Later, another snake came carrying an herb in its mouth and placed the herb on the head of the dead snake, whereupon both snakes fled. And so Asclepius used the same herb and brought Glaucus back to life. Thus the snake was placed under Asclepius's safekeeping and also among the stars. Instructed by this event, Asclepius's successors handed down the knowledge to others, so that physicians are skilled in the use of snakes.

3.13

The figure has one star on his head; one star on each shoulder; three on his left hand; four on his right hand; two on the back; one on each knee; one on the right leg; one on each foot, the right one being brighter. Thus there are seventeen stars in all. The serpent has two at the tip of its head; four under the head close together [ι Ser, ρ Ser, κ Ser, π Ser]; two on the left hand of Ophiuchus [μ Ser, υ Ser?], the one nearer the body of Ophiuchus being brighter; five on the back of the serpent [υ Ser, ν Ser, ξ Ser, ο Ser, ζ Ser], where it abuts the human figure; four at the first curve of the tail [δ Ser, λ Ser, α Ser, ε Ser]; and six stars at the second curve [μ Ser, ?, ?, ?, ?, ?], facing the third. Thus the total of stars is twenty-three.

Commentary

Ophiuchus was usually identified with Asclepius by both Greek and Roman authors, but, as noted by Hyginus, other identifications were not lacking, among them Carnabon, Triopas, Hercules and Phorbas. The stories of Carnabon and Triopas reveal a similar structure: each of the protagonists commited an offense against the goddess Ceres [Demeter], who sent a serpent to harass the offender during his lifetime, but caused him to be placed among the stars after his death. The identification of this constellation figure with Hercules is not surprising, given his status in Greek myth as the foremost slayer of monsters. Hyginus is the earliest classical source for the story of Phorbas.[1]

According to tradition, Asclepius was the son of Apollo and Coronis and was reared by Chiron, who instructed him in the art of healing. Either from Athena, who gave him the blood of the Gorgon, or from observing a snake revive its dead mate by means of a miraculous herb, Asclepius learned how to raise the dead. The association of Asclepius with a snake is consistent with ancient belief in the wisdom and healing power of serpents. Asclepius himself often appeared in the form of a serpent. The healing powers of Asclepius and his ability to revive the dead are mentioned early in Greek literature. Several ancient authors give lists of the dead men who were restored to life by him.[2]

Asclepius's power to raise the dead was the cause of his own destruction. The divine wrath was aroused, either because Asclepius was guilty of avarice, or because he acted contrary to the will of the gods, or because the diminishing death rate brought about by his success as a healer threatened to rob Hades of his due. The death of Asclepius by lightning is mentioned by several Greek authors; his subsequent elevation to the heavens, prompted by the intercession of Apollo, is a later accretion to the story.[3]

The number of stars in the figure of the man is seventeen according to Ps-Eratosthenes, Hyginus, and Hipparchus and twenty-four according to Ptolemy. The number of stars in the serpent is not given by *The Constellations*, as there is a lacuna in the text at this point. According to Hyginus, the number is twenty-three. Ptolemy lists eighteen stars in the Serpent.[4]

Orion

The Constellations 32

Hesiod says that Orion was the son of Euryale, the daughter of Minos, and of Poseidon and that to him was granted the ability to walk on the sea as he did on the land. Orion came to Chios and, while drunk, violated Merope, the daughter of Oenopion. Learning of the insult, Oenopion took it badly; he blinded Orion and banished him from the island. In his wanderings, Orion came to Lemnos and there met Hephaestus, who took pity on him and gave Orion his own servant, Cedalion, to guide him. Orion took Cedalion on his shoulders and, following Cedalion's directions, made his way to the East, where he met Helius, who cured his blindness. Orion then returned to inflict punishment on Oenopion, but Oenopion was hidden by the towns-people beneath the earth. Giving up the search, Orion went to Crete, where he hunted wild animals in the company of Artemis and Leto. He reportedly threatened to slay every beast that lived on the earth, whereupon Gaea ["Earth"] became angered and brought forth a scorpion of great size, by whose sting Orion perished. Zeus, at the request of Artemis and Leto, placed Orion among the stars to commemorate his bravery; he also set the scorpion among the stars as a reminder of the event. Others say that Orion became enamored of

Artemis when he reached manhood, and that Artemis sent against him
the scorpion by which he was stung and died, also that the gods took
pity on Orion and changed both him and the scorpion into constellations to commemorate the event.

Orion has three faint stars on the head [λ, ?, ?]; one bright star on
each shoulder [α, γ]; one faint star on the right elbow [μ]; one faint star
at the edge of the hand [ξ]; three stars on the belt [δ, ε, ζ]; three bright
stars on the sword [η, 42, θ²]; one bright star on each knee [κ, ?]; also
one bright star on each foot [β, 29]. The total is seventeen.

Poetic Astronomy 2.34

Hesiod says this is Neptune's son by Euryale, the daughter of Minos,
and that to him was granted the ability to run on the water as he did
on land, just as Iphicles reportedly had the power to run over growing
grain without bruising it.

Aristomachus, however, says there was at Thebes a certain Hyrieus
(Pindar says he lived on the island of Chios), who received Jupiter and
Mercury as his guests, and sought from them the gift of becoming a
father. Further, in order to obtain his request more easily, he sacrificed
an ox and placed it before them at a banquet. When Hyrieus had done
this, Jupiter and Mercury ordered that the hide of the ox be removed
and that the oxhide, into which they urinated, should be buried. From
the oxhide was later born a lad whom Hyrieus called Urion ["urine-
born"], because of his origin, but long-standing custom calls him
Orion. It is reported that Orion went from Thebes to Chios and there,
his desire aroused through wine, raped Merope, the daughter of
Oenopion. For this deed, he was blinded by Oenopion and driven
from the island. He went to Vulcan on Lemnos and received from him
a certain Cedalion as guide. Orion placed Cedalion on his shoulders,
and, in this way, reportedly came to Sol ["Sun"], who restored his
sight. Orion then returned to Chios to avenge himself, but Oenopion
was hidden by his countrymen beneath the earth. Giving up hope of
finding him, Orion came to the island of Crete, where he took up
hunting with Diana and, boasting to her of what we mentioned earlier,

thus came to be among the stars.

Some, however, say that Orion lived with Oenopion in excessive intimacy and, because he wished to prove his prowess in hunting, boasted to Diana of what we recounted earlier and was slain. Others, agreeing with Callimachus, say that when he tried to force himself on Diana, she shot him with her arrows, and that, on account of his hunting, he was depicted among the stars in the same pursuit. Istrus, however, says that Orion was loved by Diana and that she almost became his wife. Apollo became distressed, but his frequent scolding had no effect. On one occasion, when he observed the head of the swimming Orion from afar, he wagered with Diana that she could not hit the dark spot visible in the ocean with her arrow. She, desirous of being called the most skilled archer, shot her arrow and pierced the head of Orion. When the waves carried his body ashore, Diana grieved greatly that she had shot him and, lamenting his death with much weeping, reportedly placed him among the stars. What Diana did after his death we will relate in the stories about her.

3.33

Orion has three bright stars on his head; one star on each shoulder; one faint star on the right elbow; a like one on the hand; three on his belt; three faint stars where his sword is represented; one bright star on each knee; one on each foot. The total is seventeen.

Commentary

There are two traditions concerning Orion's parentage. According to one, he was of gigantic size and born of the Earth; according to the other, he was the son of Euryale and Poseidon. Hyginus recounts both traditions; *The Constellations* recounts only the latter, according to which Orion acquired from his father the ability to walk on the sea.[1]

The various accounts presented of Orion's death reflect a familiar theme in Greek myth: any mortal who is guilty of hubris, i.e., who boasts of his prowess or otherwise challenges the gods, incurs divine retribution (*nemesis*). Examples include women as well as men: Niobe, Arachne, Tantalus, Sisyphus.[2]

It would appear that Orion, regardless of his later connections, was originally a Boeotian hero.[3] The stories of his birth are mostly of Boeotian origin; his grave-site was shown at Tanagra, where there was a hero-cult connected with him.

The blindness of Orion occurs on the island of Chios, home of the poet Homer, who, according to tradition, was also blind.[4]

Certain of the stories connected with Orion may be astral myths, i.e., myths that arose from the relative position in the sky of two or more constellations, or of one constellation and its position relative to the horizon. Such myths include the flight of the Bears from Orion; the flight of the Pleiades from Orion; the love of Eos for Orion; Orion's ability to walk on the sea; Orion's flight from the Scorpion.[5]

Orion is one of the five constellations mentioned by Homer. The Babylonian *Tablet of the Thirty Stars* refers to the asterism of the "Mighty-Man" which has been identified with either α Orionis alone, or with the northern part of the Greek constellation.[6]

Ps-Eratosthenes, Hyginus, and Hipparchus include seventeen stars in this constellation; Ptolemy lists thirty-eight.

Pegasus

The Constellations 18

Only the front part of the Horse [Pegasus] is visible, down to the navel. Aratus says this is the horse that created the spring on Mount Helicon with its hoof, whence the spring is called Hippocrene ["equine spring"]. Others say it is Pegasus, who flew up to the stars after Bellerophon's fall. Some think that the Horse cannot be Pegasus because the constellation figure has no wings. Euripides says in his *Melanippe* that this constellation is Chiron's daughter, Hippe, who was tricked and molested by Aeolus and fled to the mountains because of her pregnancy. Her father came searching for her as she was about to give birth. She prayed that she might assume another shape so as not to be recognized when she was found, and was thus changed into a horse. Because of her own piety and also that of her father, she was placed among the stars by Artemis in a place where she is not visible to the Centaur, for that constellation is said to be Chiron. Her lower parts are not visible so that she might not be recognized as a woman.

The Horse has two faint stars on the nostril [ϵ, ν?]; one star on the head [θ]; one on the jaw [ν?]; one faint star on each ear [?, ?]; four on the neck [ρ, σ, ζ, ξ], of which the one closest to the head is the brightest; one on the shoulder [β], one on the chest [λ?]; one on the back [α]; one bright star on the edge of the belly [α And]; two stars on the front knees [η, ι]; one on either hoof [π, κ]. The total is eighteen.

Poetic Astronomy 2.18

Aratus and numerous others say this is Pegasus, the son of Neptune and Medusa the Gorgon, who created a spring on Mount Helicon in Boeotia by striking the rock with his hoof; the spring was called Hippocrene after him.

Others say that at the time Bellerophon came to Proetus, king of the Argives and son of Abas, the king's wife, Antia, fell in love with their guest and sought his attentions, promising him her husband's kingdom. When she could not persuade him, fearing he would incriminate her to her husband, she seized the initiative and told Proetus that Bellerophon tried to overpower her. Proetus, because he was fond of Bellerophon, refused to inflict punishment himself, but knowing he had a horse, sent him to Iobates, the father of Antia—whom some call Stheneboea—in order that Iobates, defending his daughter's honor, would expose Bellerophon to the Chimaera, which at that time was laying waste the fields of Lycia with flames. Bellerophon returned victorious, and, after creation of the fountain, tried to fly up to heaven—and was not far from his destination—when he looked down at the earth, and stricken with fear, fell to his death. The horse, however, flew on and was placed among the stars by Jupiter.

Others say that Bellerophon fled from Argos not under accusation by Antia, but in order not to hear what he did not wish to hear or not to be affected by her importuning. Euripides in his *Melanippe* says this is Hippe, the daughter of Chiron the Centaur, and that earlier she was called Thetis. As a young maiden on Mount Pelion she devoted herself to hunting, and at a certain time was made pregnant by Aeolus, the son of Hellen and grandson of Jupiter. As her day of delivery drew near, she fled into the forest so that her father, who thought her to be a virgin, might not see that she had begotten him a grandchild. And thus, when her father sought her, she reportedly petitioned the gods that her father not see her in childbirth. And, by will of the gods, after she gave birth, she was changed into a horse and placed among the stars.

Some say that Hippe was a seer, and was changed into a horse because she made men privy to the counsels of the gods. Callimachus, however, says that because she ceased to hunt and to worship Diana, the goddess changed her into the shape we mentioned earlier. It is said that for this reason she is not visible to Centaurus, whom many identify with Chiron; furthermore, only half of her is visible, because she did not wish it known that she was a woman.

3.17

The figure has two faint stars on the nostril; one on the head; one on the chin; one on each ear; four faint stars on the neck, of which the one nearest the head is brightest; one bright star on the shoulder; one on the chest; one on the back; one noticeable star on the navel which is called the Head of Andromeda; one on each knee; one on each hoof. The total of stars is eighteen.

Commentary

Ps-Eratosthenes differentiates between the horse which created the spring of Hippocrene on Mount Helicon and Pegasus; most ancient sources identify the two, recording that the spring was created by the hoof of Pegasus. *The Constellations* argues that this constellation cannot be Pegasus because the horse represented in the constellation is not winged. Although artistic representations of this constellation invariably show a winged horse, the literary tradition identifying the constellation as a winged horse begins with Ptolemy.[1]

Mount Helicon is situated in Boeotia between Lake Copais and the Gulf of Corinth. The eastern side of the mountain, where Hippocrene was located, was particularly sacred to the Muses. Another spring called Hippocrene, also said to have been created by Pegasus, is found at Troezen, in the eastern Peloponnesus. Pegasus was also believed to have created the spring called Pirene at Corinth.[2]

With the possible exception of Hesiod, ancient sources represent Pegasus as a winged horse. Poseidon, in the form of a horse or a bird, consorted with Medusa. Later, Pegasus and Chrysaor were said to have sprung from the blood of Medusa after she was slain by Perseus. Upon his birth, Pegasus straightway flew up to heaven where he carried the thunder and lightning of Zeus. He came down to earth once, to aid the hero Bellerophon in slaying the Chimaera, a fire-breathing monster with a lion's head, goat's body, and serpent's tail. Following that feat, Pegasus was placed in the heavens as a constellation.[3]

Bellerophon was the son of the Corinthian king Glaucus and Eurymede. After murdering his brother, Bellerophon went to King Proetus of Argos to be purified. Bellerophon's journey to the court of a king for the purpose of being cleansed of blood guilt is reminiscent of similar journeys by other Greek heroes, most notably Heracles, and reflects the priestly function of ancient Greek kings. Like other Greek heroes, Bellerophon was assisted by one of the Olympian gods, in this case Athena, in accomplishing a seemingly impossible feat. However, unlike most Greek heroes, Bellerophon was punished by Zeus for seeking, of his own accord, to join the gods by riding Pegasus to heaven. In the usual account, Zeus sent a gadfly to sting Pegasus, causing the beast to rear, and Bellerophon fell to earth and suffered lameness or blindness. The false accusation against Bellerophon by the wife of Proetus is a folk motif found in other Greek myths, e.g., Phaedra and Hippolytus.[4]

Melanippe was the child Hippe bore to Aeolus. Melanippe was the subject of two plays by Euripides, *Melanippe the Wise* and *Melanippe in Bonds*; only a plot summary survives for the former, while one long passage survives from the latter.[5]

Although the figure of a winged horse appears frequently in Ancient Near Eastern art, the figure was apparently not included among the constellations. The constellation of the Horse is probably of Phoenician origin.[6]

The Horse is comprised of eighteen stars according to Ps-Eratosthenes, Hyginus, and Hipparchus, twenty according to Ptolemy.

Perseus

The Constellations 22

It is said that this constellation was placed among the stars because of Perseus's fame. Zeus came to Danae in the form of golden rain and fathered Perseus by her. When Perseus was sent against the Gorgons by Polydectes, Hermes gave him his own helmet and sandals, with which he was able to fly. Hephaestus is said to have given Perseus a diamond wallet. Aeschylus the tragedian says in his *Phorcides* that the Gorgons were guarded by the Graeae. These had one eye among them and shared it by turns. Perseus waited until the moment of the exchange, then snatched the eye away and threw it into Lake Tritonis. Next, he attacked the Gorgons while they slept and beheaded Medusa. Thereafter, Athena wore the head of Medusa on her breast, and she granted to Perseus a place among the stars, where he is visible holding the Gorgon's head.

The figure has one star on the head [τ]; one bright star on each shoulder [γ, θ]; one bright star at the edge of the right hand [χ?]; one star on the elbow [η]; one at the edge of the left hand [Galactic Clusters 884, 869], in which hand he appears to be holding the Gorgon's head, on which there is one star [β]; one on the body [ι?]; one bright star on the right hip [α]; one bright star on the right thigh [48?]; one star on the knee [λ?]; one on the leg [μ]; one faint star on the foot [ζ]; one star on the left thigh [ν]; one on the knee [ε]; two on the leg [o, ξ]; three in the hair of the Gorgon [ω?, ρ?, π?]. The total is nineteen. The head has eight stars [β, ω, ρ, π, 16, ?, ?, ?]. The wallet has five stars but appears to have none; to some it seems visible as a nebula.

Poetic Astronomy 2.12

Perseus was placed among the stars because of his courage and because he was conceived in an unusual manner. He was sent by Polydectes, the son of Magnes, against the Gorgons. But Mercury, who loved him greatly, gave him winged sandals, a cap, and, in particular, a helmet which rendered him invisible to his opponents. The Greeks said the helmet belonged to Hades ["the invisible one"]; Perseus did not utilize the helmet of Orcus, as some claim in ignorance; no educated person could find this interpretation acceptable.

Perseus was said to have received from Vulcan a knife made of adamant, with which he killed the Gorgon Medusa—a deed no one has described. As the tragedian Aeschylus says in his *Phorcides*, the Graeae were guardians of the Gorgons; we wrote about this in the first book of the *Genealogies*. These reportedly had one eye among them, each of them using it in turn as they carried out their vigil. At the moment one of them was handing the eye over, Perseus snatched it and threw in into Lake Tritonis. Thus with the guards rendered blind, Perseus easily slew the Gorgon as she slept. Minerva placed the head of the Gorgon on her breastplate. A certain Euhemerus says the Gorgon was slain by Minerva; we will say more about her later.

3.11

He has one star on each shoulder; one bright one on the right hand, with which he is said to be holding the knife he used to kill the Gorgon; another on his left hand, with which he holds the head of the Gorgon. In addition, there is one on the left elbow; one star on the belly; another on the groin; one on the right thigh; one on the knee; one on the leg; one faint one on the foot; one on the left upper thigh; one on the knee; two on the leg; four on the left hand which are called the Gorgon's head. The total is nineteen stars. His head and the knife are not marked by stars.

Commentary

Like all the Greek heroes, Perseus was the son of a god. Acrisius, the king of Argos, had received a prophecy that he would meet death at the hands of his grandson. He therefore imprisoned his daughter, Danae, in a subterranean chamber to prevent her seeing any man. Zeus, however, reached Danae in the form of golden rain and she gave birth to Perseus. Upon their discovery by Acrisius, mother and child were set afloat in a chest. The chest washed ashore at Seriphos where a fisherman, Dictys, took Danae and Perseus under his protection. Dictys's brother, Polydectes, the king of Seriphos, wished to marry Danae but she would not consent. When Perseus grew to manhood, Polydectes sent him to bring back the head of Medusa, hoping that Perseus would die in the attempt and that Danae would then have no choice but to become his wife. But Perseus accomplished his mission, with the aid of the gods, and returned to avenge himself on Polydectes and to take Danae away with him to Argos.[1]

Medusa was one of the Gorgons, daughters of Phorkys and Ceto who lived somewhere in the West. Their appearance was so terrifying that anyone who looked at them was turned to stone. They were girded with serpents and had wings and brazen claws. Of the three Gorgons, Medusa was mortal, the other two immortal. The Gorgons were guarded by the Graeae, also daughters of Phorkys and Ceto. These were old women from birth. They were two or three in number and had only one eye and one tooth among them, which they shared in turns.[2]

Perseus's slaying of Medusa was accomplished with the aid of several gods. Ps-Eratosthenes and Hyginus mention Hermes and Hephaestus. According to other authors, Athena contributed a shield, which Perseus used as a mirror: he backed up to Medusa while she slept and cut off her head. Returning to Seriphos, Perseus caused Polydectes to be turned to stone by showing him the Gorgon's head. According to another tradition, Perseus received a shield or mirror from Athena and a sickle from Hermes, but the sandals, wallet and helmet he received from a certain nymph, whom he reached by consulting the Graeae. The version of the story in Ps-Eratosthenes and Hyginus appears to be a blending of the two traditions.[3]

In classical literature, this constellation is said to represent Perseus holding the Gorgon's head, the latter being identified with the star Algol (β Persei). The constellation itself may be of Phoenician or Egyptian origin. To the Babylonians, this constellation represented a male figure named Sugi ("the old man"), and not a hero.[4]

The number of stars comprising Perseus is nineteen according to Ps-Eratosthenes, Hyginus, and Hipparchus, twenty-nine according to Ptolemy.

Pisces

The Constellations 21

These fish are descendants of the Great Fish, whose story we shall relate more fully when we come to his constellation [38]. These fish do not lie close together. One is called the Northern Fish, the other the Southern Fish. They are connected as far as the front foot of the Ram.

The Northern Fish has twelve stars [β, γ, 7, θ, ι, κ, λ, ω, 27, 29, 30, 33], the cord has twelve stars [41, 51, δ, ε, ζ, 80, ο, π, η, ρ, μ, α], and the Southern Fish has fifteen stars [82, τ, 68, 67, 65, ψ¹, ψ², χ, υ, φ, ψ³, ?]. The cord by which the two fish are connected has three stars on the northern side, three on the southern side, three toward the east and three at the knot. The total is twelve. The total number of stars comprising the two Fish and the cord is thirty-nine.

Poetic Astronomy 2.30

Diognetus of Erythrea says that at one time Venus came with her son Cupid to the Euphrates River in Syria. Suddenly Typhon, about whom we have already spoken, appeared in the same place, and Venus and her son leaped into the river and changed themselves into fish. In this way they escaped danger. And so the Syrians who live closest to this area abstain from fish, because they are afraid to catch them, lest for similar reason they appear either to disdain the protection of the gods, or to entrap the gods themselves.

Eratosthenes says they are offspring of the Fish that we will speak about later [2.41].

3.29

These fish are connected by several stars, like a little thread, up to the front foot of Aries. The [southern] one has seventeen stars; the northern one has twelve in all. The thread connecting them has three stars facing north and three south, three facing east, and three on the knot. The total is twelve.

Commentary

The *Constellations* provides no mythological explanation for the origin of Pisces: the two fish of this constellation are said to have been placed in the heavens by virtue of their relationship to the Great Fish (Piscis Austrinus). Classical sources associate two stories with this constellation; however, neither of the two appears to be of Greek origin. The one story identifies the fish of this constellation with the two fish that found an egg of extraordinary size in the Euphrates River. They rolled the egg to dry land where a dove hatched it, and the Dea Syria emerged. The other story is recounted by Hyginus.[1]

This constellation probably originated in the fish constellation which was the twelfth sign of the Babylonian zodiac. In both Greek and Near Eastern art, the Fish are almost always represented as linked by a cord at the mouth or at the tail.[2]

The number of stars in this constellation is thirty-nine according to Ps-Eratosthenes, forty according to Hipparchus, forty-one according to Hyginus, thirty-four according to Ptolemy.

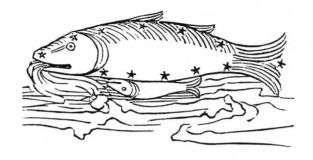

Piscis Austrinus

The Constellations 38

This is the so-called Great Fish, which is swallowing the water being poured out by the Water-Pourer [Aquarius]. Concerning this Fish, Ctesias recounts that it lived in a certain lake near Bambyce. When Derceto, who is considered a goddess by those inhabiting the region of Syria, fell into the lake one night, the Fish is believed to have rescued her. The two fish [Pisces] are said to be offspring of the Great Fish. All the fish were honored and placed among the stars because Derceto was the daughter of Aphrodite. The inhabitants of that region of Syria fashion images of fish out of gold and silver and, from the occurrence, honor them as sacred.

The Fish has twelve stars [α, β, γ, δ, ε, ζ, η, θ, ι, λ, μ, γ Gru], among them three bright ones on the nostrils.

Poetic Astronomy 2.41

The figure appears to be drinking water from the constellation Aquarius. The Fish was believed to have saved Isis when she was in distress. For this service, she placed it and its descendants, about whom we spoke earlier [2.30], among the stars. For this reason, many Syrians abstain from fish and pay tribute to gilded images of fish as household gods. Ctesias, also, writes about this.

There are twelve stars in all.

Commentary

Bambyce was one of several names given to the city which lay on the trade route between Antioch and Mesopotamia, about twenty-four miles west of the Euphrates River. Bambyce was, apparently, a Greek name, the native Syrian name being Mabog. However, in the time of Seleucus Nicator (d. 280 B.C.E.), the city was named Hieropolis and is best known by that name, which also appears on its coinage. Bambyce was the center of worship of the goddess Derceto, and boasted of a famous temple of the goddess. Near the temple was a lake in which sacred fish were kept.[1]

Derceto was the name, at Bambyce, of that Mother Earth Goddess who appeared under various names throughout the ancient Near East. A comparison of the Babylonian goddess Nanai, the Chaldean Ba'alat, the North Semitic Ishtar, the Babylonian-Assyrian Belit and the West Semitic Astarte will attest to the essential oneness of their nature. All these goddesses were consorts of the chief male divinity of their region; they were associated with the protection of cities, with war, fertility, and the moon. Temple-prostitution was common in their worship, as were eunuch priests. Aphrodite was the Greek—and considerably toned-down—version of this Eastern Mother-Earth Goddess, hence her association with Derceto in this story.[2]

Doves and fish were especially associated with the West Semitic Mother Earth Goddess. The Syrian tabu on eating fish is frequently attested in Greek literature: the Syrians believed that eating fish would cause the body to break out in ulcers and the feet and stomach to swell. Doves also were forbidden as food to the Syrians, since the Syrian Goddess was born from an egg hatched by doves. Another reason for the latter tabu may have been that Semiramis, the daughter of the Syrian Goddess, was changed into a dove.[3]

The story of Derceto's rescue by the fish is the only myth connected with this constellation in classical literature. There can be little doubt that the myth was introduced into Greece from the East, probably Syria or Phoenicia, although the constellation itself may be of either Phoenician or Babylonian origin. The mythological connection between the fish of this constellation and those of the zodiacal sign of Pisces may well reflect a more basic connection between the two constellations: the reduplication of one by the other.[4]

Ps-Eratosthenes, Hyginus, Hipparchus and Ptolemy all describe this constellation as being comprised of twelve stars.

The Planets

The Constellations 43

The five stars called "wanderers," because they have a motion peculiar to themselves, are associated with five gods. The first star is a large one, Phaenon ["shining"], associated with Zeus. The second star is not large. It is called Phaethon ["radiant"] and is named for Helius. The third star is the star of Ares. It is called Pyroeides ["fiery"]; it is not large and its color is similar to that in the Eagle. The fourth star is Phosphorus ["light-bearer"], white in color, the star of Aphrodite. This is the largest of all stars and is called by two names, Phosphorus and Hesperus. The fifth is the star of Hermes, Stilbon, small and bright. It was given to Hermes because he ordained the arrangement of the heavens, the positions of the stars, the calculation of the seasons, and the appearance of weather-signs. This star is called Stilbon ["shining"] because of its appearance.

Poetic Astronomy 2.42

It now remains for us to consider the five stars which the Greeks called *planetas* ["wanderers"]. One of these is the star of Jupiter, Phaethon by name. Heracleides of Pontus says that Prometheus, at the time he was creating man, made Phaethon so exceptional in physical beauty, that he sought to hide him and not to hand him over to Jupiter as he had done with the rest of mankind. Cupid, however, revealed Prometheus's plan to Jupiter, who sent Mercury to Phaethon to persuade him to come to Jupiter and become immortal. And thus, Phaethon was placed among the stars.

The second star is called the star of Sol; others call it Saturn's star. Eratosthenes says this star is named after Sol's son, Phaethon. Many have written about him, how he was unwittingly carried away by his father's chariot and scorched the earth, and how Jupiter struck him with a lightning bolt so that he fell into the Eridanus River and was placed among the stars by Sol.

The third star is that of Mars, which others call Hercules's star. This star follows that of Venus, as Eratosthenes explains, for the reason that when Vulcan wed Venus, and she, because of his watching her, was unable to grant her favors to Mars, Mars requested of Venus that only his own star should be visible following her star. And thus, because love burned within him, he called his star Pyroides ["fiery"].

The fourth star is that of Venus, Lucifer ["light-bearer"] by name; some say it is Juno's star. But there is a tradition, recounted in many stories, that its name is Hesperus. Of all stars, this appears to be the largest. Some say this is the son of Aurora and Cephalus, who surpassed many men in beauty. For that reason, he was said to rival even Venus, as Eratosthenes recounts, and is called the star of Venus, appearing both at sunrise and at sunset. Thus, as we said earlier, he is rightly called both Eous ["dawn star"] and Hesperus ["evening star"].

The fifth star is that of Mercury, Stilbon by name. It is small and bright. It was reportedly granted to Mercury because he was the first to institute the system of months and to ordain the movement of the stars. Euhemerus, however, says that it was Venus who first instituted the order of the stars and revealed it to Mercury.

Commentary

The ancient Greeks appear to have distinguished the planets from the other stars in the fifth century B.C.E., although literary evidence suggests that Venus was known, as a regular star, in the time of Homer. At first, the Greek called the planets *planetes asteres* ("wandering stars"), having observed that their motion differed from that of the other stars. The name *planetae* ("planets") first occurs in the fourth century B.C.E.[1]

The planet Venus was called *eosphorus* ("dawn-bearer") and *hesperus* ("evening star") by Homer, depending on whether it appeared at dawn or at dusk, and was apparently regarded as two different stars. The identification of the morning and evening stars was attributed by several ancient authors to the Pythagoreans.[2]

According to ancient sources, the Greeks learned to distinguish the five planets from the other stars, and also adopted the names of the planets, from the Babylonians, who identified each of the five planets as the star of a particular deity. Thus, the star of Marduk, the chief god of the Babylonian pantheon, was assigned to Zeus by the Greeks, while the star of the mother-goddess Ishtar was assigned to Aphrodite, the star of Nabu, the son of Marduk and god of sciences, letters and scribes, to Hermes, and the star of the war deity, Nergal, became the star of Ares. Less obvious is the assignment of the star of Ninurtu, a fertility and war god, to Cronus. The identification of the Greek with the Babylonian names is supposed to have been made by the Pythagoreans. Subsequently, the names of other Greek deities were substituted for or added to the names assigned by the Pythagoreans. Thus, the star of Ares was also called the star of Heracles, on the basis of a syncretistic identification of the two figures; the star of Aphrodite was also called the star of Hera because its Babylonian name, Ishtar-Baltis, meant "queen of heaven." According to Cumont, the confusion ensuing from this overabundance of names caused the Greek astronomers of the later Alexandrian period to coin a new set of names based on the appearance of the planets, for the purpose of establishing a fixed terminology. It was at that time that the names Phaenon, Phaethon, Pyroeis/Pyroeides, Phosphorus and Stilbon were introduced.[3]

Ps-Eratosthenes assigns the planets to the various deities on the basis of the Pythagorean identifications, except for the star of Cronus, which he assigns to Helius; he also gives the later names based on the appearance of the planets. Hyginus gives all the names associated with the planets as well as the epithets applied to each one. In addition, he provides a myth to explain the origin of each planet; no such myths are recounted in *The Constellations*.[4]

The arrangement of the heavenly phenomena is attributed to Hermes both in this passage and in the passage referring to the constellation Triangulum. Aratus attributes the arrangement of the stars to Zeus. In the Babylonian creation epic, as well, the arrangement of the celestial phenomena is attributed to the chief god, Marduk.[5]

The Pleiades

The Constellations 23

The Pleiades are located on the nape of the Bull's neck, and consist of seven stars, said to be the daughters of Atlas. Hence, this constellation is called "seven-starred." Only six of the seven stars are visible; the reason given for this is the following. It is believed that six of the sisters consorted with gods, and one with a mortal. Three lay with Zeus: Electra, the mother of Dardanus; Maia, the mother of Hermes; and Taygete, the mother of Lacedaemon. Two lay with Poseidon: Alcyone, the mother of Hyrieus, and Celaeno, the mother of Lycus. Sterope reportedly lay with Ares and bore Oenomaus. Merope, however, lay with Sisyphus, a mortal, and so appears very faint. The Pleiades are held in the highest honor among men, as they appear at all times of the year. According to Hipparchus, this constellation is prominent, having a triangular shape.

Poetic Astronomy 2.21

Alexander says the Hyades were thus called because they were daughters of Hyas and Boeotia, and the Pleiades because they were daughters of Pleione—whose father was Oceanus—and of Atlas. The Pleiades are seven in number, but only six are discernible. The reason given for this is that six of the seven consorted with immortals—three with Jupiter, two with Neptune, one with Mars—in contrast with the remaining one who was the wife of Sisyphus. Jupiter was the father of Dardanus by Electra, of Mercury by Maia, of Lacedaemon by Taygete; Neptune was the father of Hyreus by Alcyone, Lycus and Nycteus by Celaeno. Mars was the father of Oenomaus by Sterope, who, according to others, was the wife of Oenomaus. Merope married Sisyphus and bore Glaucus, who many say was the father of Bellerophon. Merope was placed among the stars because of her sisters; however, because she married a mortal, her star is faint. Others say that the faint star is

Electra, who, after Troy was captured and her descendants (through
her son Dardanus) were driven into exile, removed herself out of grief
from the Pleiades—who are believed to lead the circular motion of the
stars—and withdrew to the arctic circle, where, with hair loosed, she
has long been observed in mourning. For this reason she was called
Cometes ["long-haired"].

The ancient astronomers configured the Pleiades outside the constellation Taurus. As we mentioned earlier, these were daughters of
Pleione and Atlas. Now, Pleione was traveling through Boeotia with
her daughters when Orion was aroused and desired to possess her, but
she fled. Orion chased after her for seven years but was unable to find
her. Jupiter, pitying the daughters, placed them among the stars. They
were later called "the Bull's Tail" by many astronomers. And even now,
Orion appears to follow them as they set, in their flight; our writers
called these stars Vergiliae ["spring stars"] because they rise after the
vernal equinox; indeed, they enjoy greater honor than other stars,
because the rising of their sign signals summer, while its setting signals
winter. No other signs are accorded this role.

Commentary

The use of this constellation as a weather sign by the ancients is
attested by both Greek and Roman authors. Hesiod, e.g., advises the
farmer to begin his harvest with the rising of the Pleiades and his
ploughing with their setting, while the sailor is advised to haul his boat
onto the land when the Pleiades set and the seas become stormy.[1]

Although Ps-Eratosthenes offers no myth to explain the origin of
this constellation, other Greek authors provide several accounts, one
of which is recounted by Hyginus. One tradition not mentioned by
Hyginus recounts that the seven maidens, who were companions of
Artemis, were pursued by Orion until Zeus, in answer to the maidens'
pleas, placed both pursuer and pursued in the heavens, where the chase
continues. According to another tradition, the seven maidens were
placed in the heavens on account of their grief over their father's plight.

A third tradition connects the Pleiades with the Hyades, noting that all were sisters of Hyas, or alternatively, nurses of Dionysus. Still another tradition reports that the Pleiades were the doves which carried ambrosia to the infant Zeus.[2]

Ps-Eratosthenes and Hyginus recount that the Pleiades are so called from the name of their mother, Pleione. Other ancient sources propose additional derivations, e.g., from Gk. *plein* ("to sail")—because the heliacal rising of these stars in May marked the beginning of the ancient sailing season—or because these stars are useful to "many" (Gk. *pleious*) on land and sea, or because the maidens changed themselves into "doves" (Gk. *peleias*) while fleeing Orion.[3]

There were several ancient traditions concerning the "invisible" Pleiad. Ps-Eratosthenes and Hyginus recount that she was either Merope or Electra. According to other accounts, one of the Pleiades (who is unnamed) was struck by lightning and thus rendered invisible.[4]

The Euphratean constellation known as "The Clusterers" may be the antecedent of the Greek constellation.[5]

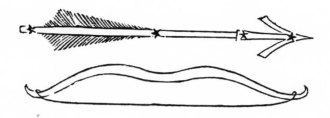

Sagitta

The Constellations 29

    This figure is said to be the arrow of Apollo. Angered at the fate of Asclepius, Apollo reportedly used this arrow to kill the Cyclopes, who fashioned the thunderbolt of Zeus. Apollo hid the arrow among the Hyperboreans, where the feather temple, too, is located. He is said to have retrieved the arrow later, when Zeus freed him and ended his service to Admetus, about which Euripides speaks in the *Alcestis*. It is believed that the arrow was then brought back through the air with fruit-bearing Demeter. The arrow was of great size, as Heraclides of Pontus says in his treatise *On Justice*. Apollo changed the arrow into a constellation and placed it among the stars to commemorate his struggle.

    The Arrow has one star on the tip [γ]; one faint star on the shaft [ζ?]; two stars on the notch [α, β]. One of these stars is very bright. The total is four.

Poetic Astronomy 2.15

    This is said to be the arrow with which Hercules slew the eagle that was consuming the liver of Prometheus; many have seen fit to write about this. Now the ancients carried out their sacrifices to the immortals with great ceremony, and sacrificial victims were totally consumed in the ceremonial flame. When, on account of the great expense involved, sacrifices became unaffordable for the poor, Prometheus, who, because of his marvelous and supreme inventiveness, was believed to have created men, reportedly persuaded Jupiter that mankind should throw part of the sacrificial animals into the flame, and consume the other part in their meal; later custom confirmed this practice. Since he had obtained this easily, as from a god, not as from a greedy human, Prometheus himself sacrificed two bulls. He first placed their entrails on the altar, then heaped the meat from both bulls together and covered it with a bull's hide. He then gathered together the bones and covered them with the remaining hide and allowed Jupiter to choose between the two heaps. Jupiter, not acting with divine forethought, nor as a god who ought to foresee all things, was deceived by Prometheus—for we have decided to give credence to the stories—and, thinking that both heaps were meat, chose the bones for his portion. And so, in subsequent rites and religious sacrifices, after the sacrificial meat was consumed, the remains, which were the gods' portion, were burned in the same flame.

But to return to our previous theme, Jupiter, when he discovered what had happened, was angered and took fire away from mankind, so that the beneficence of Prometheus should not outweigh the power of the gods, and so that meat would be useless to mankind since it could not be cooked. Prometheus, however, accustomed to scheming, planned to snatch fire away and restore it to mankind. Thus, when the other gods were away, he came to Jupiter's flame, and hiding a small part of it in a fennel-stalk, he seemed not to run, but to fly, joyfully shaking the fennel-stalk so that the enclosed air might not extinguish the flame with its vapors in the narrow space. And so, to this day, men who are bringers of good news usually come quickly; later, following the practice of Prometheus, the custom was established that runners in athletic contests would run brandishing a torch.

Motivated by Prometheus's act to bestow an equivalent gift, Jupiter gave mankind the woman fashioned by Vulcan and endowed with all gifts by will of the gods—thus her name, Pandora ["she of all gifts"]. Prometheus, however, he bound with a chain of iron on a Scythian mountain called Caucasus; the tragedian Aeschylus says that Prometheus remained bound for thirty thousand years. In addition, Jupiter sent an eagle that continually consumed Prometheus's liver, which grew back at night. Some say this eagle was the offspring of Typhon and Echidna, others, of Terra and Tartarus. Many say that the eagle was fashioned by Vulcan and brought to life by Jupiter.

Concerning the liberation of Prometheus, the following has been handed down by tradition. When Jupiter, struck by the beauty of Thetis, pursued her and did not prevail with the timid maiden, but was no less determined to succeed, the Parcae prophesied what was to be: they said that whoever was the husband of Thetis would sire a son more glorious than his father. Prometheus, awake not by desire but by necessity, heard this and reported it to Jupiter. The latter, fearing that which he himself had done to his father Saturn, gave up his desire to make Thetis his wife—so as not to be deprived of his patrimony—and, in return for the service of Prometheus, freed him from his chains; however, because he had sworn an oath, Jupiter did not release Prometheus from bonds altogether. He ordered Prometheus, as a memorial, to bind his finger with the two elements of stone and iron. Men took up the custom of wearing rings fashioned of stone and iron as a tribute to Prometheus. Some say Prometheus wore a wreath, as if to proclaim himself a victor who had sinned with impunity. And so, men established the custom of wearing wreaths in times of great joy and in victory; this is seen in athletic contests and at banquets.

But to return to the beginning of the story and the slaying of the eagle: Hercules, sent by Eurystheus in search of the golden apples, did not know the way and came to Prometheus, who, as we said earlier, was chained on Mount Caucasus. Prometheus showed him the way; and Hercules, returning victorious, desired both to tell Prometheus about the slaying of the dragon, concerning which we spoke earlier [2.3], and to thank him for the service rendered. Hercules quickly attacked the eagle that had been feeding on the liver of Prometheus and dispatched it with an arrow. For this reason, when animals were sacrificed, men decided to offer livers on the altars of the gods, in order to appease the latter for the liver of Prometheus.

Concerning the arrow, Eratosthenes says it was the one used by Apollo to slay the Cyclopes who fashioned the lightning bolt of Jupiter. (Many say that Jupiter used the lightning bolt to slay Asclepius.) Apollo buried this arrow on the Hyperborean mountain; but when Jupiter pardoned him, this same arrow was carried back to Apollo by the wind, along with grains that grew at that time. Thus, the arrow was placed among the stars.

This figure has four stars in all, of which one is at the beginning of the shaft, the second in the middle, and the two remaining ones at the spot where the iron tip is attached; these appear discretely.

Commentary

Hyginus follows Ps-Eratosthenes in identifying this constellation with the arrow of Apollo. An alternative explanation, namely, that this was the arrow used by Hercules to rescue Prometheus, provides Hyginus with the opportunity for a long digression on Prometheus, which includes several aetiological stories related to him: the tricking of Zeus by Prometheus, which explains why the ancient Greeks offered only the inedible parts of a sacrificial animal to the gods; the restoration of fire to the human race by Prometheus; the origin of victory crowns; the custom of running to deliver good news; and the offering of livers as a sacrifice.[1]

If the origin of this constellation is Greek and based on its resemblance to an actual arrow, as suggested by several scholars, then it is not surprising that the constellation would be associated with two preeminent archers of Greek myth, Apollo and Heracles, and with two of their best-known feats: the slaying of the Cyclopes by Apollo, and of the eagle that fed on Prometheus's liver by Heracles.[2]

The death of the Cyclopes at the hands of Apollo was one of a series of related events beginning with the slaying of Asclepius by Zeus and leading to the events described in Euripides's *Alcestis*. Apollo slew the Cyclopes, who fashioned the thunderbolts by which Asclepius died, but was himself punished by being forced to pass a year in the service of a mortal, Admetus, the king of Thessaly. Concerning the Cyclopes, there are several traditions. In Homer, the Cyclopes were described as a race of giant shepherds whose manner of life was uncivilized: they had no laws, each followed his own will, and they paid no homage to the gods.[3] In Hesiod, the name Cyclopes was used of three sons of Uranus and Gaea: Arges, Steropes, and Brontes. These were one-eyed creatures who were cast into the underworld on two occasions, once by their father Uranus, and once by Cronus, with whom they sided to overthrow Uranus. After their second imprisonment they were released by Zeus, and were instrumental in Zeus's victory over the Titans, providing him with thunderbolts and lightning. Thereafter, they were retained by Zeus to fashion his thunderbolts. In later authors, the Cyclopes also appeared as assistants of Hephaestus or as a tribe of skillful builders, credited with building the "Cyclopaean walls" of Argos, Tiryns, and Mycenae.[4]

Most ancient sources agree that Apollo suffered exile and bondage in consequence of his slaying the Cyclopes. This sentence reflected actual Athenian practice with respect to a man who killed another either without design or with some good cause. In such cases, the homicide was required to leave his country by a particular road and to remain in exile for a specified length of time, usually eight years. At the end of his exile, the homicide was purified by sacrifices and expiatory rites. It is not surprising that all such rites of purification were considered to be under the charge of Apollo.[5]

The earliest extant reference to the Hyperboreans occurs in the *Homeric Hymn to Dionysus*, although Herodotus states that they were mentioned in Hesiod and other early poets. The Hyperboreans, as their name implies, lived "beyond the north wind." Herodotus placed their land beyond the north slope of the Altai Mountains. Living, so to speak, "behind" the north wind, the Hyperboreans enjoyed a serene climate, which was reflected in their character and manner of life.

They were a pious people, associated especially with the worship of Apollo, and extremely long-lived—they lived for one thousand years according to some sources. Their diet included no meat, and they suffered from no sickness or disease. Roman sources relate that when they became tired of their easy life, they leapt from a cliff into the sea.[6]

Apollo's connection with the Hyperboreans is first attested in Greek literature by Pindar. A much earlier connection is suggested by tradition, according to which Tempe, Delphi and Delos, all three well-known centers of Apolline worship, were either founded by the Hyperboreans or closely connected with them. There are two accounts of the Hyperborean connection with Delphi: the local poetess Boeo (fourth century B.C.E.) stated that the Delphic oracle was founded by the Hyperboreans, and a fragment of Alcaeus (seventh–sixth century B.C.E.) recounts that Apollo, shortly after his birth, was ordered by Zeus to fly to Delphi in a swan-drawn chariot to establish an oracle. Instead, Apollo flew to the land of the Hyperboreans where he remained for one year, establishing laws. The people of Delphi, meanwhile, continuously prayed to Apollo to come to them. The god finally arrived among the Delphians in mid-summer, bringing sheafs of grain. His arrival was thereafter celebrated every ninth year at midsummer. Herodotus records the Delian tradition, according to which two Hyperborean maidens came to Delos at the same time as Apollo and Artemis. In commemoration of this fact, the Hyperboreans sent annual offerings to Delos via Scythia and Greece. These offerings were wrapped in wheat-straw and were not brought to Delos by the Hyperboreans in person, but were passed along from one nation to another until they reached their destination. According to Herodotus, the gifts of the Hyperboreans arrived regularly in his day. Hyperborean offerings were also brought to Delphi, by whom it is not certain, along the sacred road which passed through Tempe.[7]

The temple of feathers mentioned in *The Constellations* is attested by several late Greek sources, but all these point to earlier tradition, which held that the temple of feathers was built at Delphi by bees from bees' wax and feathers. This was the second temple to be built at

Delphi, the first having been established by the Hyperboreans. After its completion, the feather temple was flown to the land of the Hyperboreans by Apollo.[8]

The treatise *On Justice* by Heraclides Ponticus was a dialogue recording a discussion between Abaris the Hyperborean and Pythagoras. Abaris appeared in Greek literature as a religious teacher and prophet, and was often mentioned in connection with such legendary figures as Aristeas of Proconnesus, Epimenides of Crete, Pythagoras, and Empedocles. According to most ancient sources, Abaris traveled about Greece carrying Apollo's arrow as a badge. Heraclides, however, said that Abaris traveled about by means of Apollo's arrow.[9]

The origin of this constellation is uncertain. There is no evidence to suggest an Ancient Near Eastern arrow-constellation. Some modern scholars have argued in favor of a Greek origin based on the resemblance of the constellation to an arrow.[10]

The number of stars comprising the Arrow is four according to Ps-Eratosthenes, Hyginus, and Hipparchus, five according to Ptolemy.

Sagittarius

The Constellations 28

This figure represents the Archer [Sagittarius], whom most believe to be Centaurus. Others say it is not Centaurus because the Archer does not appear to have four legs, but to be standing and shooting a bow—and no centaur used a bow. The Archer represents a man with the legs and tail of a horse, like the satyrs. For this reason, many think it impossible that the Archer is Centaurus and believe that the figure represented is Crotus, the son of Eupheme, who was the nurse of the Muses. Crotus dwelled and subsisted on Mount Helicon. According to Sositheus, he invented archery and was subsequently inspired by the Muses to obtain his food from wild animals. Crotus lived in the company of the Muses, and as he listened to them, he signified his approbation by a sound, marking time to their rhythmless song by a clapping of his hands. Others imitated him, and the Muses, having gained glory by his good-will, asked Zeus to grant distinction to Crotus as he was a pious person. Thus, Crotus was placed among the stars because of the use of his hands, taking his archery along with him as an attribute. His deed remained among men. The Ship [Argo] bears witness to him, for he is visible not only to those on land, but to those at sea as well. Those, therefore, who write that this constellation is

Centaurus are in error.

The Archer has two stars on the head [ξ^2, o?]; two on the bow [ϵ?, λ?]; two on the tip of the arrow [γ, ϕ]; one on the right elbow [51]; one on the edge of the hand [δ]; one bright star on the body [σ]; two stars on the back [τ?, ψ]; one on the tail [ω?]; one on the front knee [α]; one on the hoof [$\beta^{1,2}$?]; one on the back knee [ι]. The total is fifteen. The remaining seven stars are under the leg; these are similar to those on the back and are not at all bright.

Poetic Astronomy 2.27

Many say this is Centaurus; others, however, say this cannot be for the reason that no centaur used arrows. The reason is sought why he is represented with a horse's limbs and has a tail like the satyrs. Many say this is Crotus, the son of Eupheme, who was the nurse of the Muses. Sositheus the tragedian says that Crotus lived on Mount Helicon and delighted in the company of the Muses, also that he was an avid hunter. His diligence in those pursuits won him great acclaim, for he was both the swiftest in the forest and the most accomplished in the musical arts. Because of his accomplishment, the Muses asked Jupiter to represent him among the stars, and Jupiter acquiesced. Since he wished to signal all of Crotus's skills in one figure, Jupiter gave him a horse's legs, because he used horses extensively; he added arrows so that by virtue of these Crotus's skill and speed might be evident; and he placed a satyr's tail on his body because the Muses delighted in him no less than Liber delighted in the satyrs. In front of his feet are a few stars which form a circle; some say these form his wreath, cast off as if by one at play.

3.26

He has two stars on the head; two on the bow; two on the arrow; one on the right elbow; one in the front hand; one on the belly; two on the back; one on the tail; one on the front knee; one on the foot; one on the back knee. There are fifteen in all. The wreath of Centaurus consists of seven stars.

Commentary

The centaurs are more abundantly represented in Greek art than in literature. Part-man, part-horse creatures descended from Ixion and Nephele, they were a wild, lustful race who inhabited Mount Pelion in Thessaly. In archaic art, the centaurs were represented as complete men with the barrel and hind legs of a horse attached to their back. In later art, the figures became more equine and were represented as horses with the body of a man from the waist up. The statement in *The Constellations* that no centaur used a bow is at first glance surprising, but apparently true. The character of the centaurs may have survived in the popular imagination of Modern Greece as the *kallikantzaroi*, the mischievous, slightly naughty creatures who make their annual appearance during the fortnight following Christmas.[1]

The name Centaurus was often used to designate the best known of the centaurs, Chiron. However, Chiron was not a typical centaur. Not only was his genealogy different—he was the son of Cronus and Philyra, and was part-horse, part-man because his father assumed the shape of a stallion to consort with his mother—but also, unlike the other centaurs, Chiron was quiet and peace-loving. He had a reputation for wisdom and justice and many Greek heroes were his pupils. He was well-versed in music and medicine. In art, too, Chiron was represented as more human in form than the other centaurs: he retained the earlier form of the centaurs (a complete man with the barrel and hind legs of a horse), and was usually shown wearing a cloak. Furthermore, Chiron was immortal while the other centaurs were mortal.[2]

The satyrs were human-shaped creatures with horse's tails and often goat's legs or horns. They may have symbolized the fertility of woods and hills—the Oreads (i.e., nymphs of the mountains) were their sisters. Like the centaurs, the satyrs were wild, lustful, and fond of revelry. They were often represented as followers of Dionysus. Similarities in the character, and to some extent in the appearance, of the satyrs and the centaurs have led some scholars to posit a common origin for these two groups of creatures.[3]

Just as Chiron represented a more subdued and civilized aspect of the centaurs, so Crotus was more subdued and civilized than the other satyrs. Crotus was associated not with the ribald followers of Dionysus, but with the Muses. Like the other satyrs, Crotus was the son of Pan and Eupheme, the nurse of the Muses. The tradition attributing to Crotus the invention of archery and rhythmic clapping occurs only in astronomical literature.[4]

Centaurus and Crotus are the only two figures associated with this constellation by classical authors. The implication in *The Constellations* that the identification of the Archer as Centaurus was widespread is supported by scant literary evidence; in astronomical literature, the Archer is usually identified as Crotus. Ps-Eratosthenes argues for the identification of the Archer as Crotus rather than Centaurus, on the basis of the figure's appearance. He might also have mentioned that Centaurus (Chiron) was already represented by his own constellation, the constellation Centaurus. Despite the objection posed by Ps-Eratosthenes to the identification of this constellation figure as Centaurus, it is likely that both a two- and a four-legged creature were associated with Sagittarius in Greece at any given time.[5]

The origin of the Archer can be traced back to the Euphratean half-man, half-horse constellation figure which was always represented as shooting a bow. A similar figure is found among the Egyptian constellations.[6]

The number of stars in Sagittarius is fifteen according to Ps-Eratosthenes and Hyginus, sixteen according to Hipparchus, thirty-one according to Ptolemy.

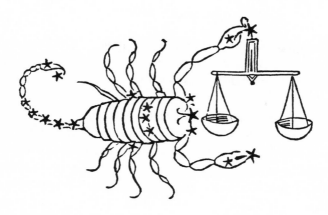

Scorpio and Libra

The Constellations 7

Scorpio extends over two twelfths of the zodiac because of its size. One part comprises the claws [Libra], the other the body and tail. It is related that Artemis brought a scorpion forth from the hill on the island of Chios and that it stung Orion because he had unbecomingly used force against her during a hunt; also that Orion died from its sting. Zeus placed the scorpion among the bright stars so that succeeding generations of men might know its strength and power.

Scorpio has two stars on each claw [α Lib, μ Lib, β Lib, δ Lib], of which those in front are bright, those behind faint; three stars on the head [β, δ, π], of which the middle one is brightest; three bright stars on the back [σ, α, τ]; two stars on the belly [13?, d?]; five on the tail [ε, μ^{1,2}, ζ^2, η, θ]; and two on the stinger [λ, υ]. The westernmost star, and the most brilliant of all, is the bright star on the northern claw [α]. The total is nineteen.

Poetic Astronomy 2.26

This figure is divided into two signs because of the size of its constituent parts, one of which our writers call Libra. The entire constellation appears to have been created because Orion, priding himself on his supreme prowess as a hunter, boasted to Diana and Latona that he could slay any creature arising from the earth. Whereupon, it is recounted that Earth, in anger, sent a scorpion to kill him. Jupiter, admiring the fortitude of both combatants, placed the scorpion among the stars, so that the sight of it might serve as a lesson to men, lest any one of them should grow too self-confident. However, Diana, because of her care for Orion, petitioned Jupiter to grant her the same honor he had freely bestowed on Earth; and so Orion was placed among the stars in such a way that when Scorpio rises, Orion sets.

3.25

This figure has two stars on each of the so-called Claws, of which the foremost are brighter. In addition, there are three stars on the face, of which the middle one is very bright; three on the back; two on the belly; five on the tail; two on the tip of the tail, with which the scorpion is said to sting. Thus, the total number of stars is nineteen.

Commentary

According to Greek myth, the immediate cause of Orion's death was either Artemis herself or a scorpion. In the earliest version, Artemis shot Orion because he was loved by the goddess Eos ("Dawn.") In subsequent versions Orion was shot by Artemis because he tried to rape the goddess or one of her maidens during a hunt, or else Artemis or Gaea sent the scorpion to kill Orion because he boasted of his prowess as a hunter, or because he attempted to use force against Artemis. The story was set by various authors on Chios, Delos or Crete.[1]

Both Hyginus and Ps-Eratosthenes identify this figure with the scorpion that caused Orion's death. The myth of Orion's death by the scorpion's bite may well be of astral origin: when the constellation Scorpio rises, that of Orion sets, creating the impression that the scorpion is pursuing Orion, while Orion is fleeing the scorpion.[2]

The constellation Scorpio appears to be of Babylonian origin. The Babylonians distinguished various parts of the Scorpion—claws, head, body, tail, sting—but considered the whole one constellation. The Greeks at first pictured the Scorpion as one constellation, but later depicted the Claws as a separate constellation. The first mention of Scorpio in Greek literature can be traced to Cleostratus of Tenedos (*c.* 520 B.C.E.).[3]

The number of stars in Scorpio is nineteen according to Ps-Eratosthenes and Hyginus, seventeen according to Hipparchus, and twenty-one according to Ptolemy.

According to *The Constellations*, the brightest star in the constellation Scorpio is the "bright star on the northern claw" (α Lib). Ptolemy, writing *c.* 150 C.E., attributes the same magnitude to this star and to Antares (α Sco). At present, the magnitude of α Lib is 2.7, while that of Antares is 1.2. It would appear either that α Lib has decreased in brightness since classical antiquity, or that Antares has increased.[4]

Taurus

The Constellations 14

According to Euripides's *Phrixus*, the Bull [Taurus] was placed among the stars because he carried Europa from Phoenicia over the sea to Crete. Because of this service, he was honored by Zeus among the brightest stars. Others say the Bull represents Io and that the constellation was honored by Zeus on her account.

The stars called Hyades make up the forehead and face of the Bull. Toward the edge of the backside are the Pleiades, consisting of seven stars. The Pleiades are called seven-starred for this reason. Only six of the stars are visible; the seventh is extremely faint.

The Bull's head is comprised of seven stars. He is moving backward and has his head turned toward his body. At the base of each horn he has one star [τ, 97], of which the one on the left is brighter. There is one star on each eye [α, ε]; one on the nostril [γ]; one on the tip of each horn [β, ζ]. The above stars are called the Hyades. In addition, there is one star on the left front knee [90]; one on each hoof [88, ν]; one on the right knee [μ]; two on the neck [37, ω]; three on the back [φ?, χ?, ψ?], the last one of which is bright; one under the belly [o?]; one bright star on the chest [λ]. The total is eighteen.

Poetic Astronomy 2.21

The Bull is said to have been placed among the stars because he carried Europa safely to Crete, as Euripides recounts. Many say that when Io was changed into a cow, Jupiter, in order to please her, placed her among the stars. Her front part resembles a bull, but the rest of the body is fainter. It faces the rising sun.

The stars comprising the face of Taurus are called Hyades. Pherecydes the Athenian says these were the nurses of Liber, seven in number, who were formerly called Dodonian nymphs. Their names are as follows: Ambrosia, Eudora, Pedile, Coronis, Polyxo, Phyto, Thyone. It is said they were chased by Lycurgus, and all except Ambrosia fled to Thetis, as Asclepiades recounts. But Pherecydes says they brought Liber to Thebes and entrusted him to Ino. For this reason, Jupiter in gratitude placed them among the stars.

They are called Hyades, for, as Musaeus says, there were fifteen daughters born to Atlas and Aethra, the daughter of Oceanus. Of these, five were called Hyades because Hyas their brother was much beloved by his sisters. When he was killed by a lion while hunting, five of the above-mentioned sisters reportedly perished because of their unending grief. And so, those who were most affected by his death were called Hyades. The remaining ten sisters discussed their sisters' death and seven of them killed themselves; wherefore, as many will have realized, they were called Pleiades.

3.20

Taurus has a star on each horn, the one on the left being brighter; one on each eye; one in the middle of his face; two in that spot where the horns arise. Concerning the seven stars called Hyades, many say the two stars we spoke of last are not part of the constellation, but that there are only five Hyades in all. In addition, there is one star on the left front knee and two over the hoof; one on the right knee and two on the curve of the neck; three on the upper back, of which the hindermost is brightest; one under the belly; one on the chest. The total, excluding the Vergiliae, is eighteen.

Commentary

The story of Europa and the bull is recounted with no major variations by several Greek and Latin authors. Zeus fell in love with Europa, the daughter of Agenor, the king of Tyre, and approached her in the form of a white bull. Europa was so taken with the beauty and gentleness of the creature that she climbed onto its back and was carried away over the sea to Crete. The bull was usually held to be Zeus himself, but in a few versions the bull was a real bull, sent by Zeus to carry off Europa.[1]

The myths of Europa, of Pasiphae's passion for the bull sent by Poseidon, and of Theseus and the Minotaur all attest to the prominence of bulls in the legends of Crete. Archaeological evidence points to the importance of bulls in Cretan religion, as well.[2]

Io was the daughter of the river-god Inachus in most versions of her story. She was loved by Zeus then changed into a heifer either by him, to hide her from Hera, or by Hera herself as punishment. The connection of Hera with cows was a long-standing one, and Io was said in most accounts to be a priestess of Hera.[3]

Ancient sources relate that the bull sent to Minos by Poseidon or Zeus was later captured by Heracles and finally slain by Theseus at Marathon.[4]

According to one tradition, the constellation of the Hyades represents the sisters of Hyas, who were placed among the stars by Zeus when they died of grief over the death of their brother. The Pleiades are sometimes said to be sisters of the Hyades, who also died of grief and became stars, but later than the Hyades. According to another tradition, the Hyades were the nymphs who reared Dionysus, either at Dodona or at Nysa, and were changed by him into stars. Some ancient sources derive the name Hyades from the Y-shape of the constellation or from the association of the constellation with rain (Gk. *hyetos*).[5]

The constellation Taurus appears in the heavens as the front half of a bull or cow with unusually long horns, moving backward across the sky. A bull-constellation is found in both Babylonian and Egyptian records as the first sign of the zodiac. The reason for this position at the head of the zodiacal constellations was the fact that Taurus coincided with the vernal equinox between the years 4000 and 1800 B.C.E.[6]

The number of stars in Taurus is eighteen according to Ps-Eratosthenes, Hyginus, and Hipparchus, thirty-three according to Ptolemy, who includes the Pleiades in this number.[7]

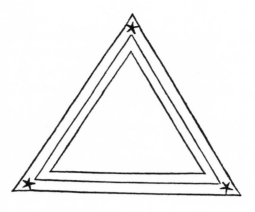

Triangulum

The Constellations 20

This constellation lies above the head of the Ram [Aries]. It is believed that the easily distinguishable letter taken from the beginning of Zeus's name was placed in this position by Hermes, who arranged the stars in the heavens, because the Ram is a faint constellation. Some say that the geographical configuration of Egypt is modeled on the Delta-shape in the heavens [Triangulum] and that the Nile River caused the outline of Egypt to be such, while, at the same time, providing safety for the land, rendering the earth more receptive of seed, and tempering the climate for the harvest of crops.

The Delta-shape has three stars [α, β, γ], all of them bright, one at each corner.

Poetic Astronomy 2.19

This constellation is called "Delta-shape" because it is triangular, like the Greek letter delta [Δ]. Mercury is said to have placed it above the head of Aries, so that by its brilliance it should mark the place where Aries shines faintly, and also that it might form the first letter of Jupiter's name in Greek. Others say it represents the position of Egypt; others of Ethiopia and Egypt, where the Nile marks their boundary; others think it represents Sicily; others say there are three angles because the gods divided the world into three parts.

Commentary

There is no aetiological myth explaining the origin of this constellation. The constellation is called the Triangle by some writers of astronomical literature, the Delta-shape by others.[1]

Both Ps-Eratosthenes and Hyginus note that Triangulum represents either the first letter of Zeus's name in Greek, placed above the Ram to make the latter more conspicuous, or some aspect of Mediterranean geography.[2]

Ps-Eratosthenes suggests that Egypt itself is a reflection of the constellation. This is noteworthy, considering that other constellation figures are said to be the result of some condition or action on earth. However, in ancient Egyptian religious thought the land of Egypt itself was believed to be an image of the heavens. Thus, unless the Greek text is corrupt, it would appear that *The Constellations* is referring to an Egyptian theme.[3]

Geometric shapes are not uncommon among the Babylonian constellations, and it is likely that the Greek constellation of the Delta-shape was of Babylonian origin.[4]

The constellation consists of three stars according to Ps-Eratosthenes, Hyginus, and Hipparchus; four according to Ptolemy.

Ursa Major

The Constellations 1

Hesiod says this constellation is Lycaon's daughter, who lived in Arcadia and chose to lead a hunter's life in the mountains with Artemis. Ravished by Zeus, she remained with Artemis, but kept her condition from the goddess. Later, she was seen while bathing by Artemis and discovered to be near childbirth, whereupon the goddess grew angry and changed her into a wild beast. And so, in the shape of a bear, the maiden gave birth to the child called Arcas. While in the mountains, she was caught by some goatherds and handed over to Lycaon along with her child. After a time, being unaware of the law prohibiting it, she entered the sacred precinct of Zeus. Pursued by her own son and the Arcadians, she was about to be put to death for breaking the aforementioned law, when Zeus, mindful of their previous relationship, rescued her and placed her among the stars. He named her the Bear [Ursa] because of what had befallen her.

This figure has seven faint stars on the head [o, 2, π^2?, ρ, σ^2, τ, 23]; two stars on each ear [24, ?, ?, ?]; one bright star on the shoulder blade [?]; one star on the chest [υ]; two on the front leg [ι, κ]; one bright star on the back [α]; one bright star on the belly [β]; two stars on the back legs [γ, δ]; two on the paw [λ, ν]; three on the tail [ϵ, ζ, η]. The total is twenty-four.

Poetic Astronomy 2.1

Hesiod says this figure is Callisto, the daughter of Lycaon, who reigned in Arcadia. Because of her love of hunting, Callisto joined the company of Diana, to whom she was very dear because of their similar character. Later, ravished by Jupiter, she feared to tell Diana what had befallen her. But she was unable to conceal her condition very long, for when she was near the day of delivery, heavy with child and weary with exertion, she bathed her body in the river, and Diana knew she had not preserved her virginity. Because of her greatly misplaced trust, the goddess repaid Callisto with no small punishment: her maidenly form was stripped away and she was changed into a bear, which in Greek is called *arktos*. In this form, Callisto gave birth to Arcas. But Amphis the comic poet recounts that Jupiter, disguising himself as Diana, pursued the maiden as she hunted, as if to assist her, then, when she was removed from the sight of her companions, ravished her. When Diana inquired of her what had happened that she appeared to be with child, Callisto replied that what had occurred was Diana's fault. And thus, Diana changed her into the form mentioned earlier.

Wandering in the forest as a wild animal, Callisto was captured by a group of Aetolians and brought, with her son, to Lycaon in Arcadia as a gift. There it is said that, being ignorant of the law, she wandered into the temple of Lycaean Jupiter, followed closely by her son. When the Arcadians, in pursuit, attempted to slay them, Jupiter, mindful of his indiscretion, snatched Callisto and her son away and placed them among the stars. He named her Arctus, and her son, of whom we will speak later [2.4], he named Arctophylax ["Bear-keeper"].

Some say that after Callisto was ravished by Jupiter, Juno became angry and changed her into a bear, which came within view of Diana as she hunted and was killed by her, then later recognized and placed among the stars. But others say that when Jupiter was pursuing Callisto into the forest, Juno, suspecting the course of events, was determined to say she had apprehended him in the act. Jupiter, in order more easily to conceal his indiscretion, changed Callisto into a bear and abandoned her; Juno, finding a bear in place of a maiden, pointed the bear out to Diana, who was hunting, so that she might kill the animal. When Jupiter perceived what had happened, he was angered and placed the likeness of a bear among the stars.

This constellation, as many note, does not set and those who wish to find a reason for this say that Tethys, the wife of Oceanus, would not accept this constellation to set in the sea with the others, because Tethys had been the nurse of Juno, whom Callisto, as Jupiter's paramour, had supplanted.

The historian Araethus of Tegea says the maiden was called not Callisto, but Megisto, and was the daughter not of Lycaon, but of Ceteus, and so Lycaon's granddaughter; he says, furthermore, that at an earlier time The Kneeler [Hercules] was called Ceteus. In other respects the story is similar to what we recounted earlier; these events were supposed to have occurred on Mount Nonacris in Arcadia.

The figure has seven stars on the head, all faint; two on each ear; one bright star on the shoulder; two on the front foot; one at the top of the back; two on the foremost of the two back legs; two on the back foot; three on the tail. All the stars of the constellation are twenty-two.

Commentary

The Constellations does not mention the name of the maiden who was changed into Ursa Major in the present narrative. It is clear, however, both from Hyginus, who cites the same source as Ps-Eratosthenes, as well as from the myth recounted in connection with Ursa Minor, that her name was Callisto. The story of Callisto is probably of Arcadian origin. Not only is Arcadia the locale of the story recounted here, but archaeological and literary evidence attests that Artemis was worshiped throughout Arcadia, often with the epithet Calliste ("most lovely"). Additional Arcadian elements in this story are the metamorphosis of Callisto into a bear, an animal sacred to Artemis, and the fact that Callisto is said to be the mother of Arcas, the eponymous ancestor of the Arcadians.[1]

The association of Callisto with Zeus Lycaeus in this story provides a further tie with Arcadia. Mount Lycaeum is located in Arcadia, and the Zeus worshipped there in antiquity had some connection either with wolves or with light. An altar of Zeus Lycaeus surrounded by a sacred precinct was situated at the summit of Mount Lycaeum. According to Pausanias, entrance into the sacred precinct was forbidden, and anyone who broke the tabu, whether man or beast, lost his shadow and either died or was put to death within a year.[2]

The literary tradition of this story begins with Hesiod, who is cited as a source by both Ps-Eratosthenes and Hyginus. However, it is unlikely that the changing of Callisto into a constellation was mentioned by Hesiod. The association of Callisto with the constellation Ursa Major, at least in literature, appears to date from the third century B.C.E.[3]

Three names are applied to this constellation figure in Greek literature: *arktos* ("bear"), *hamaxa* ("wagon"), and Helice. The first name is the most common and associates the constellation figure with the story of Callisto. The name "wagon" is used of this constellation by Homer and Aratus. As Helice this constellation figure is identified with one of the two nurses of Zeus (the other was transformed into Ursa Minor), whom he placed among the stars out of gratitude for their service. All three of the above names are applied to both Ursa Major and Ursa Minor.[4]

The widespread identification of these stars with a bear figure among native peoples inhabiting both North America and Siberia leads one scholar to suggest that this constellation may date from the time before the sea engulfed the land bridge between the two continents, about 15,000 years ago, forming the Bering Strait.[5]

Ps-Eratosthenes lists twenty-four stars in this constellation; Hyginus lists twenty-two. Ptolemy lists twenty-seven.[6]

Ursa Minor

The Constellations 2

This is the Bear called "small"; most call her Phoenice. She was honored by Artemis, who, learning that Zeus had violated Phoenice, changed her into a wild animal. Later, after Phoenice was rescued, it is said that Artemis honored her by placing a second image among the stars, so that she received double honors. Aglaosthenes, in his *Naxica*, says that Zeus had a nurse, Cynosura, who was one of the Idaean nymphs. Her name occurs in the city called Histoe, which was founded by the followers of Nicostratus. There, the harbor and its surrounding area were named Cynosura after her. Aratus calls this figure Helice and says that she was from Crete, also that she was the nurse of Zeus and was therefore adjudged worthy of honor in heaven.

The Small Bear has one bright star at each corner of the square [β, γ, ζ, η], and three bright stars on the tail [α, δ, ε], making a total of seven. Beneath the second of the two westernmost stars, there is another star, called Polus [α], around which the entire universe appears to revolve.

Poetic Astronomy 2.2

Aglaosthenes, who wrote the *Naxica*, says this is Cynosura, one of the Idaean nymphs, who were nurses of Jupiter. The city of Histoe, which was founded by Nicostratus and his followers, its port, and the greater part of its region are called Cynosura, after her. She was among the Curetes who were attendants of Jupiter.

Some say the nymphs Helice and Cynosura were nurses of Jupiter, and for that reason, as a reward for their service, were placed in the firmament and both called Arctus ["bear"]; we call them *Septentriones* ["seven oxen"]. But many say that Ursa Major resembles a wagon, and, indeed, the Greeks called that constellation *hamaxa* ["wagon"]. This is the reason for its commemoration. Those who first observed the heavens and assigned the stars to a particular figure named this constellation the Wagon, not the Bear, because two of its seven stars, being similar and very close together, are said to be oxen, while the remaining five resemble a wagon. For this reason, those who assigned names wished to call the sign closest to that constellation Bootes ["ox-driver"]. We will speak more about him later [2.4].

Aratus, however, does not explain the names of Bootes and the Wagon in this way. He says it was because Ursa Major appears to turn like a wagon around the north pole and Bootes to drive her. In this he errs greatly. For later on, as Parmeniscus explains, some astronomers increased the number of seven stars to twenty-five so that the bear's figure should not consist solely of seven stars. And accordingly, the figure that was formerly called Bootes because he followed the wagon was now called Arctophylax ["Bear-keeper"], and in Homer's time, the "seven oxen" were called Bear. Homer says that the constellation is called both Bear and Wagon, but nowhere does he say that Bootes was called "Bear-keeper."

Many wonder why Ursa Minor is called "Phoenician" and why those who sail by that constellation are said to navigate more precisely and carefully, and why, if Ursa Minor is more precise than Ursa Major, everyone does not sail by the former. Those people do not know the reason for the appellation "Phoenician." Now, Thales of Miletus, who was diligent in researching these matters and was the first to call that constellation "Bear," was from Phoenicia, as Herodotus states. And so, those who inhabit the Peloponnesus sail by Ursa Major, but the Phoenicians sail by the constellation they received from its discoverer and, following it carefully, are believed to navigate more precisely, and indeed call the constellation "Phoenician" because of its discoverer.

3.1

The figure has one bright star at each corner, and three above the tail, seven in all. Among the westernmost stars of the tail the last one is called Polus, as Eratosthenes recounts. Around that star the firmament is believed to revolve. The remaining two stars are called *choreutae* ["choral dancers"] because they circle around Polus.

Commentary

Two traditions are evident in the stories concerning Ursa Minor. In one tradition, Ursa Minor represents a nurse of Zeus whom he placed among the stars out of gratitude. Often, that nurse is mentioned together with another nurse, who is changed into Ursa Major. In the other tradition, the story of Ursa Major is transfered to Ursa Minor and the latter is identified with a maiden who suffers the same fate as Callisto.[1]

Hyginus notes that both Ursa Major and Ursa Minor were nurses of Zeus and that the Greeks applied the names "Bear" or "Wagon" to Ursa Major, and the name "Bear" to Ursa Minor, while the Romans referred to both as "oxen," thus providing an explanation for the name "Bootes" ("ploughman"). He also explains why Ursa Minor was called the "Phoenician" constellation and notes that it is a more accurate navigational guide than Ursa Major.[2]

The story concerning Phoenice related by *The Constellations* contains several puzzling elements. Phoenice is "honored" by Artemis, who changes her into a wild animal, after which Phoenice is somehow "saved," whereupon Artemis places "another" image of her among the stars, so that Phoenice receives "double honors." The nature of Phoenice's salvation is not clear. Moreover, the phrases "second

image" and "double honors" are puzzling if they refer to the origin of
one constellation, as Ps-Eratosthenes seems to imply. If, however, this
story is viewed as a continuation of the story about the Greater Bear,
as Franz proposes, then *The Constellations* is referring to the origin of
two constellations (Ursa Major and Ursa Minor), and the inconsisten-
cies in the account of Ps-Eratosthenes are removed: Phoenice is to be
identified with Callisto, who was raped by Zeus, changed into a wild
animal by Artemis, and finally into the constellation of the Greater
Bear by Zeus; Artemis, seeing the honor bestowed on Phoenice/
Callisto by Zeus, hastened to "double" that honor by placing a second
image of Phoenice/Callisto (Ursa Minor) among the stars.[3]

Following its introduction into Greece *c.* 550 B.C.E., the constellation Ursa Minor was referred to by various names in classical literature, but was usually mentioned in connection with Ursa Major,
whether the two figures were associated with nymphs, wagons, or
oxen.[4]

Ps-Eratosthenes, Hyginus, Hipparchus, and Ptolemy all assign
seven stars to this constellation.

Virgo

The Constellations 9

Hesiod in the *Theogony* says this figure is Dike ["Justice"], the daughter of Zeus and Themis ["Divine Law"]. Aratus, who takes the story from Hesiod, says that Dike was immortal and formerly dwelt on earth among the human race. When humans began to change for the worse and no longer to uphold justice, she did not remain among them, but withdrew to the mountains. Then, when civil strife and wars beset the human race because of its total injustice, she despised mortals altogether and went up to heaven. There are numerous other stories concerning this constellation. Some say it is Demeter because of the sheaf of grain she holds, others say it is Isis, others Atargatis, others Tyche ["Fortune"] and for that reason they represent her as headless.

The figure has one faint star on the head [ν?]; one star on each shoulder [?, ?]; two on each wing [β, η?, ε, ρ]: the one on the right wing between the shoulder and the wingtip is called Protrygeter [ε] ["pre-harvester"]; one star on each elbow [?, ?]; one at the tip of each hand [χ?, α], of which the bright one on the left [α] is called Stachys ["sheaf"]; six faint stars along the hem of the robe [ι, κ, φ, ?, ?, ?]; one faint [. . .] one star on each foot [λ, μ]. The total is twenty.

Poetic Astronomy 2.25

Hesiod says this figure is Themis, the daughter of Jupiter; but Aratus says she is the daughter of Astraeus and Aurora, and that she reigned over the human race during the Golden Age of mankind. Because of her diligence and fairness she was called Iustitia ["Justice"]. At that time no neighboring nations were provoked to war by others, nor was any transport utilized, since the human race cultivated the fields for its livelihood. But after the death of that generation, those who were born became less dutiful and more greedy. And so Justice withdrew from among men. Finally, the calamity came to such a point that the bronze race of men was born. And she, unable to bear any more, flew up to the stars.

Others say she is Fortuna, others Ceres; between the two, the latter is less likely because her head is very obscure. Many say she is Erigone, the daughter of Icarius, about whom we spoke earlier [2.4]. Others say it is Apollo's daughter by Chrysothemis, who as an infant was called *parthenos* ["maiden"], and who, because she died young, was placed by Apollo among the stars.

3.24

There is one faint star on the head; one on each shoulder; two on each wing, of which one star which is on the right wing, at the shoulder, is called Protrygeter; and one on each elbow. In addition, there is one star on each hand, of which the one on the left hand is larger and brighter. That one is called Stachys. There are six stars scattered on her dress and one on each foot. Thus the total is nineteen stars.

Commentary

Dike and her sisters, Eunomia ("Order") and Eirene ("Peace"), collectively known as the Horae ("Seasons"), were the offspring of the allegorical union of Zeus and Themis. In recounting the story of Dike, both Ps-Eratosthenes and Hyginus evoke the myth of the Ages of Man related by Hesiod and numerous other Greek and Roman authors. The myth traces the history of mankind beginning with the Golden Age, when mankind "lived like the gods without sorrow of heart, remote and free from toil and grief." After the Golden Age, the moral deterioration of mankind from one generation to the next—symbolized in the myth by metals of decreasing value: gold, silver, bronze—resulted in the Iron Age, when all restraint and order disappeared from the Earth.[1]

As Ps-Eratosthenes correctly observes, numerous mythological figures are identified with the constellation Virgo. Between them, Ps-Eratosthenes and Hyginus provide seven different identities for this constellation. Additional identities, including Thespia, Eileithyia, Cybele, Athena, and Hecate, are provided by later sources. The identification of this constellation figure as a woman, most often a maiden, appears to have been borrowed by the Greeks from the Babylonian tradition, which associated the constellation with the maiden-aspect of the tripartite nature of the Great Mother Goddess.[2]

The number of stars comprising Virgo is twenty according to Ps-Eratosthenes, nineteen according to Hyginus and Hipparchus, and twenty-six according to Ptolemy.

Appendix 1

Corresponding Greek and Latin Names

Greek spellings are Latinized. English usage follows the Latin names unless otherwise noted.

Greek Names	Latin Names
Artemis	Diana
Ares	Mars
Aphrodite	Venus
Athena	Minerva
Cronus	Saturnus (Saturn)
Demeter	Ceres
Dionysus	Liber
Gaea	Terra (Earth)
Hades	Orcus (Hades)
Helius	Sol
Hephaestus	Vulcanus (Vulcan)
Hera	Juno
Heracles	Hercules
Hermes	Mercurius (Mercury)
Hesperus	Vesper
Leto	Latona
Nephele	Nubes
Phosphorus	Lucifer
Polydeuces	Pollux
Poseidon	Neptunus (Neptune)
Persephone	Proserpina
Zeus	Jupiter

Appendix 2

Constellation Names and Abbreviations

Andromeda	And
Aquarius	Aqr
Aquila	Aql
Ara	Ara
Argo	Pup, Vel, Car, Pyx
Aries	Ari
Auriga	Aur
Bootes	Boo
Cancer, Asini, Praesepium	Cnc
Canis Major	CMa
Canis Minor	CMi
Capricorn	Cap
Cassiopeia	Cas
Centaurus	Cen
Cepheus	Cep
Cetus	Cet
Coma Berenices	Com
Corona Borealis	CrB
Crux	Cru
Cygnus	Cyg
Delphinus	Del
Draco	Dra
Eridanus	Eri
Gemini	Gem
Grus	Gru
Hercules	Her
Hydra, Crater, Corvus	Hya, Crt, Crv
Leo	Leo
Lepus	Lep

Libra	Lib
Lupus	Lup
Lyra	Lyr
Ophiuchus	Oph
Orion	Ori
Pegasus	Peg
Perseus	Per
Pisces	Psc
Piscis Austrinus	PsA
Pleiades	Tau
Sagitta	Sge
Sagittarius	Sgr
Scorpio	Sco
Serpens	Ser
Taurus	Tau
Triangulum	Tri
Ursa Major	UMa
Ursa Minor	UMi
Virgo	Vir

Appendix 3

Star Charts

The star maps on the following two pages show the shapes and positions of the constellations as they would have appeared in the time of Ps-Eratosthenes and Hyginus. They are reproduced with some simplifications from *Eratosthenis Catasterismi cum Interpretatione Latina et Commentario*, edited by J. C. Schaubach (Göttingen, 1795).

These star maps differ in two small ways from modern charts of the constellations. First, modern star charts use coordinate lines centered on the celestial equator and poles, while these maps follow the Greek practice of using coordinate lines based on the ecliptic, the path of the sun and planets through the zodiac. Second, due to the precession of the equinoxes, the north celestial pole is now centered near Polaris (α UMi), located on the tip of the tail of Ursa Minor. In these maps, the north celestial pole is placed between the tail of Ursa Minor and Draco, which is where it was located in the time of Ps-Eratosthenes and Hyginus.

Northern Hemisphere

Southern Hemisphere

Notes

Introduction

1. See W. W. Tarn, *Hellenistic Civilization*; T. B. L. Webster, *Hellenistic Poetry and Art*.

2. G. E. R. Lloyd, *Greek Science After Aristotle*; R. E. Wycherley, *How the Greeks Built Cities*.

3. Callimachus, *Aetia*, prologue. See also D. A. Russell, *Criticism in Antiquity*, 34–41; B. H. Fowler, *The Hellenistic Aesthetic*.

4. P. E. Easterling and B. M. W. Knox (eds.), *Greek Literature*, 33.

5. For example, Nicander, *Theriaca* (a guide to noxious animals), and *Alexipharmaca* (on treating snake-bites); Numenius, *The Banquet* (on cooking), *Halieutica* (on fishing); Aratus, *Phaenomena*.

6. P. Walcot, *Hesiod and the Ancient Near East*. In contrast to epic poetry, which dealt with themes of love or war, didactic (from Gk. *didaskein*, "to instruct") literature treated themes taken from science, philosophy, arts, and crafts, with the intent of offering practical advice or instruction.

7. Strabo, *Geography* 1.2.3.

8. An *aetion* ("cause") was a story that explained the origin of a natural phenomenon, or a custom, or a place name. Strictly speaking, each of the stories in *The Constellations* and *Poetic Astronomy* is an *aetion* explaining the origin of a constellation. Some of those stories include further examples of aetiological stories, e.g.: the Athenian custom of suspending swings from trees during a certain festival recalled the fate of Erigone (p. 58); the region of Cynosura was named after the nymph of that name (pp. 201–2). Callimachus himself composed the *Aetia*, a catalogue of myths related to cult-origins; the *Ctiseis* by Apollonius of Rhodes was a collection of foundation legends associated with various cities of the ancient world.

9. The fragments of Eratosthenes's writings have been studied by G.

Bernhardy, *Eratosthenica*; H. Berger, *Die Geographischen Fragmente des Eratosthenes*; P. J. Parsons and H. Lloyd-Jones, *Supplementum Hellenisticum*, 397–99; J. Powell, *Collectanea Alexandrina*, 58–68.

10. F. Solmsen, "Eratosthenes' *Erigone*: a Reconstruction," *Transactions of the American Philological Association*, 78 (1947): 252–75. See also pp. 56–58.

11. Longinus, *On the Sublime* 33.5.

12. The most recent study, E. Gürkoff, *Die Katasterismen des Eratosthenes* summarizes earlier findings, while a discussion of the questions posed by the text of the *Catasterismi* can be found in J. Martin, *Histoire du texte des Phénomènes d'Aratos*.

13. See W. Sale, "The Popularity of Aratus," *Classical Journal* 61 (1965): 160–64.

14. The Greek verb *dokeuo* means "to keep an eye upon," "watch narrowly."

15. See the explanation provided by Hyginus (p. 199).

16. The Greek astronomer and mathematician Hipparchus of Nicaea was born about 190 B.C.E. and made a series of astronomical observations between 161–126 B.C.E. His only surviving work is a commentary on Aratus and Eudoxus in which he notes errors of Eudoxus that were repeated by Aratus.

Andromeda

1. On Andromeda's rescue by Perseus, see Apollodorus, *The Library* 2.4.3; Ovid, *Metamorphoses* 4.665; Hyginus, *Fables* 64; Strabo, *Geography* 16.759; Pausanias, *Guide to Greece* 4.35.9. See also pp. 75–76, 83–84, 85, 157–58. Josephus, *The Jewish War* 3.9,. says that traces of Andromeda's fetters were still visible at Joppa in his day (first century C.E.). Perseus's first-born son was said to be the ancestor of the Persian kings. On the other children of Perseus and Andromeda, see Apollodorus, *The Library* 2.4.5. Perseus and Andromeda came to Argos by way of Seriphos and Tiryns. *The Constellations* is the only source to recount that Andromeda spurned her parents' plea to return to them.

2. Thompson, *Motif-Index of Folk-Literature*, R111.1.3: "Rescue of princess from dragon"; A531.4: "Hero conquers sea monster"; F628.1.0.1: "Strong man slays monster"; H335.3.1: "Princess sacrificed to dragon." The preponderance of folk-tale elements throughout the Perseus legend is untypical of Greek legendary cycles. See S. Morenz, "Die orientalische Herkunft der Perseus-Andromeda Sage," *Forschungen und Fortschritte* 36 (1962): 307–9. See also Allen, *Star Names*, 31–40.

3. On the origin of this constellation, see Allen, *Star Names*, 31–40. Allen suggests a Euphratean origin for the entire Perseus-Andromeda group. On the Phoenician name Adamath, see Brown, *Primitive Constellations*, 1:48–50.

Aquarius

1. See Scholiast on Aratus, *Phaenomena* 283; Hyginus, *Fables* 224, Nonnus, *Dionysiaca* 12.38, 105; Scholiast on Germanicus BP 68, 85.8, G 153.16; Ovid, *Fasti* 2.145.

As to the association of this constellation with water, see F. Boll and H. Gundel, "Sternbilder," in W. H. Roscher, *Ausführliches Lexikon der griechischen und römischen Mythologie*, 976. Aquarius is said to represent the spirit of the Nile (Pindar, cited by Scholiast on Aratus, *Phaenomena* 283). There are also two late identifications of doubtful validity connecting Aquarius with Hebe (Teucrus in Boll, *Sphaera*, 281; see also Pausanias, *Guide to Greece* 2.13.3, who says that at Phlius, Hebe was also called Ganymeda), and with Aristaeus, the son of Apollo and Cyrene, and lover of Eurydice (see Vergil, *Georgics* 4.453–59; Scholiast on Germanicus BP 68).

2. See p. 35.

As to the parentage of Ganymede, who was usually said to be the son of Tros and Callirhoe, see Apollodorus, *The Library* 3.12.2; Hyginus, *Fables* 271, 224. As to the abduction of Ganymede by Tantalus or Minos, see P. Friedländer, "Ganymedes," in *Paulys Realenzyklopädie der Altertumswissenschaft*, 7:737–49.

For the gift of horses to Tros, see Homer, *Iliad* 5.265; Pausanias,

Guide to Greece 5.24.5; Apollodorus, *The Library* 2.5.95. The golden vine is mentioned by Scholiast on Euripides, *Orestes* 1392.

3. See Allen, *Star Names* 45–55; Boll and Gundel, 974–77.

Aquila

1. No eagle is mentioned by Homer, *Iliad* 20.232. An eagle sent by Zeus is mentioned by Apollodorus, *The Library* 3.12.2; Vergil, *Aeneid* 5.253. The abduction of Ganymede by the eagle first appears in art during the fourth century B.C.E. See W. Drexler, "Ganymedes," in Roscher, *Lexicon*, 1:2.1598–1603. See pp. 29–31.

The tradition that Zeus abducted Ganymede out of love goes back at least to the poet Theognis (sixth century B.C.E.).

2. See pp. 125–26, where the reason given for the catasterism of the Lion is its prominence as king of beasts. The references to the eagle as the "king of birds" as well as its association with Zeus, are numerous. See D'Arcy Thompson, *A Glossary of Greek Birds*, 2–16.

3. Hyginus is the earliest extant source of the myths of Merops and Mercury-Anaplades; both myths are referred to by later authors: see Strabo, *Geography* 17.808, and Aelian, *The Nature of Animals* 13.33.

4. As is evident from the ancient texts, there were gaps between the constellations. Stars in those gaps were referred to as "unformed," and sometimes, as in the case of Antinous and Coma Berenices, served as the basis for a "new" constellation. Contemporary sky maps contain no gaps, reflecting the constellation boundaries established by international agreement in 1930. On the historical Antinous, see Pausanias, *Guide to Greece* 8.9.4; Dio Cassius, 69.11.1–4; *Historia Augusta*, 14.5; Ammianus Marcellinus, 22.16.2. The death of Antinous is usually dated to 130 C.E.

For references to the constellation Antinous in the sixteenth to eighteenth centuries, see Allen, *Star Names*, 41.

5. See Frankfort, *Cylinder Seals*, 138–39 and Plate XXIVh; Aelian, *The Nature of Animals* 12.21, connects the eagle with the Babylonian hero Gilgamesh.

Ara

1. See Hesiod, *Theogony* 617–819; Apollodorus, *The Library* 1.2.1; Horace, *Odes* 3.4.42; Hyginus, *Fables* 150; Claudian, *Battle of the Gods and the Giants* 27. See also pp. 50, 52, 61–63, and 71–72. Aratus, *Phaenomena* 403–35, explains the constellation of the Altar as an attempt by Night to provide a guide for sailors.

2. See Scholiast on Aratus, *Phaenomena* 403, 436; Scholiast on Aratus, *Phaenomena* 403, 436; pp. 79–80; Manilius, *Astronomy* 1.421; Hipparchus, 1.11.6; Ptolemy, *Almagest* 8.1.

As to Greek altar-types, see the archaeological study by C. G. Yavis, *Greek Altars*, 132, 136, 165, 171. Two older studies, based more on literary evidence, are A. de Molin, *De ara apud Graecos*, and E. Reisch, "Altar," in *Realenzyklopädie* 1:1640–91. As to Ancient Near Eastern altar-types with respect to this constellation, see Brown, *Primitive Constellations*, 1:217.

3. As to the history of the Altar, see Allen, *Star Names*, 61–64, 273; Boll and Gundel, 1016–18.

Argo

1. See Apollonius Rhodius, *The Voyage of Argo*; Orpheus, *Argonautica*; Valerius Flaccus, *Argonautica*; Ovid, *Metamorphoses* 7.1–404; Apollodorus, *The Library* 1.9.16–28; Hyginus, *Fables* 12–26.

Excavations at Iolcus have dated the site to the pre-Mycenaean (i.e., Middle Helladic) period. See P. A. Theochares, "Iolcus," *Archaeology* 11 (1958), 13–18.

The crew of the Argo was consistently numbered at fifty by ancient authors. Compare the myth of Danaus and his fifty daughters (see Apollodorus, *The Library* 2.11). Modern scholars speculate that the Argo was probably a pentekontor (=fifty-oared ship) of the type described by Homer. As to evidence for the use of pentekontors in the Mycenaean Age; see R. T. Williams, "Ships in Greek Vase Painting," *Greece and Rome* 18 (1949): 126. The pentekontor was used until 550–525 B.C.E., when it was replaced by the triremes; see L. Casson, *The Ancient Mariners*, 92. On pentekontors in general, see J. S. Morrison,

Greek Oared Ships, 7; C. Torr, *Ancient Ships,* 3.

2. As to the folk-motif of the magic ship, see Thompson, *Motif-Index,* D1123: "Magic ship"; D1610.11: "Speaking ship"; E841: "Extraordinary ship"; D1610.2.1: "Speaking oak"; D1311.4.2: "Speaking tree gives prophecy"; D1610: "Magic speaking objects"; H1332.1: "Quest for golden fleece."

3. See Aratus, *Phaenomena* 342; Scholiast on Aratus, *Phaenomena* 342, 348; Scholiast on Germanicus BP 97.13, G 172.17.

The Symplegades were situated at the entrance to the Black Sea and, in Greek myth, were a serious obstacle to ships. After the Argo succeeded in passing through them unharmed, except for the slight damage to its stern-ornament, the Symplegades ceased to clash together and remained forever open.

An artistic representation of seventh century B.C.E. Phoenician war-galleys is discussed by R. D. Barnett, "Early Shipping in the Near East," *Antiquity* 32 (1958): 226–27. See also Torr, *Ancient Ships,* plate 2, figure 10. On the shape of early Greek ships, see Williams, "Ships in Greek Vase Painting," 126; G. S. Kirk, "Ships on Geometric Vases," *Annals of the British School at Athens* 44 (1949): 93–153; Casson, *The Ancient Mariners,* passim; Morrison, *Greek Oared Ships,* 7–84. A perusal of R. O. Faulkner, "Egyptian Seagoing Ships," *Journal of Egyptian Archaeology* 26 (1940), 3–9, will show that the Egyptians, like the Greeks, built ships with curved prows. On ships of the ancient Near East, see Torr, *Ancient Ships,* passim; Casson, *The Ancient Mariners,* passim.

As to the constellations Taurus and Pegasus, see pp. 191–94 and 151–55.

4. See Aratus, *Phaenomena* 340; Scholiast on Aratus, *Phaenomena* 342, 348; Manilius, *Astronomy* 1.412; Scholiast on Germanicus BP 97.13, G 172.17; Hyginus, *Fables* 14; Plutarch, *On Isis* 22.

As to the origin of the constellation, see Allen, *Star Names,* 64–75; Boll and Gundel, 1005–8.

5. As to the star Canopus, see pp. 105–7. In modern times, this constellation has been divided into four smaller constellations, labeled

Carina ("Keel"), Puppis ("Stern"), Pyxis ("Compass"), and Vela ("Sails"), respectively.

Aries

1. See Allen, *Star Names*, 75–83. Aries is identified as the golden lamb of Thyestes and Atreus by Lucian, *Of Astrologers* 12. The constellation is said to include Athena by Teucer, first century C.E. (See F. Boll, *Sphaera*, 270). After Medusa the Gorgon was slain by Perseus (see pp. 157–58), Athena wore the Gorgon's head on her breastplate.

2. For the story of Phrixus and Helle, see Apollodorus, *The Library* 1.9.1. The account in Hyginus of Demodice's unsuccessful attempt to seduce Phrixus, and her subsequent accusation of Phrixus to her husband is a familiar motif (Hippolytus and Phaedra, Joseph and Potiphar's wife). See Thompson, *Motif-Index*, K2111: "Potiphar's wife."

3. See Apollodorus, *The Library* 1.9.1. Hyginus says that Phrixus sacrificed the ram to Zeus. Other sources mention Ares or Hermes as the recipient of the sacrifice. The ram already had a golden fleece when it rescued Phrixus and Helle, according to most accounts. According to Scholiast on the *Iliad* 7.86, the ram acquired its golden fleece subsequent to the rescue of Phrixus and Helle.

4. On Aeetes, see Hesiod, *Theogony* 956–58; Apollodorus, *The Library* 1.9.1; Homer, *Odyssey* 10.137. Helius was the Greek god of the Sun (see p. 133). Circe is best known for changing Odysseus's men into swine (Homer, *Odyssey* 10.233–43). Pasiphae was the wife of Minos and mother of the Minotaur (see pp. 87–90). Medea helped Jason to win the golden fleece and, in most accounts, became his wife (see Euripides, *Medea*).

5. As to the custom of sacrificing a king's son in time of famine, see Frazer, *Golden Bough*, 3:161–63. On the connection of such sacrifices with the family of Athamas, see Herodotus, *The Histories* 7.197. On Zeus Laphystius, see Farnell, *Cults*, 1:93–94; Cook, *Zeus*, Appendix B. Mount Laphystium is a western continuation of Mount Helicon (see

Leake, *Travels in Northern Greece*, 2:140–42).

Alus in Thessaly was founded by Athamas, according to Strabo, *Geography* 9.433. As to human sacrifice at Alus, see Herodotus, *The Histories* 7.197; Frazer, *Golden Bough*, 1:213–15. On the connection of rams with Zeus, see Cook, *Zeus*, 1:414–19, 405–9; Farnell, *Cults*, 1:95–96.

6. Apollodorus, *The Library* 1.9.1.

7. Herodotus, *The Histories* 2.54–57; Arrian, *History of Alexander* 3.34.

8. Pausanias, *Guide to Greece* 5.1.4.

Auriga

1. For the story of Erichthonius's birth, see Apollodorus, *The Library* 3.14.5; Pausanias, *Guide to Greece* 1.2.6, 14.6; Hyginus, *Fables* 166. Erichthonius succeeded Cecrops as king of Athens, or he expelled Amphictyon and became king. On Erichthonius as the inventor of the chariot, see Vergil, *Georgics* 3.113. Compare Herodotus, *Histories* 4.189, who says that the Greeks learned to yoke four horses to a chariot after they obtained Libyan horses. On the use of silver in Attica, see Pliny the Elder, *Natural History* 7.197 and Hyginus, *Fables* 274. On Erichthonius as the founder of the Panathenaean Games, see Apollodorus, *The Library* 3.14.6. According to Plutarch, *Life of Theseus* 24, the Panathenaean festival was instituted by Theseus. A sacred snake lived in the Erechtheum in the fifth century B.C.E., according to Herodotus, *Histories* 8.41.

On the *apobates*, who leapt from one horse to another during the race, and the *parabates*, see D. Demetrakou, *Mega Lexicon tes Hellenikes Glosses*.

The statue of Athena referred to in the present story is the ancient olive-wood image which stood in the Erechtheum.

2. For the story of Oenomaus, see Apollodorus, *Epitome* 2.4–9. The sea into which Myrtilus was thrown by Pelops was named the Myrtoan Sea after him.

3. As to Amalthea, see Apollodorus, *The Library* 2.7.5. See also pp.

50, 52, and 73. As to Amalthea's horn, the original *cornu copiae*, see Cook, *Zeus*, 1:501–3. As to Zeus's use of the goat's skin as armor during the Titanomachia and the epithet "aegis-bearing," see Scholiast on Aratus, *Phaenomena* 156; Scholiast on Homer, *Iliad* 2.547.

4. As to the history of this constellation, see Boll and Gundel, 915–20; Allen, *Star Names*, 83–92.

Bootes

1. In connection with the present story, see pp. 197–200. On Arcas as the son of Callisto and Zeus, see Apollodorus, *The Library* 3.8.2; Hyginus, *Fables* 155, 176, 224; Scholiast on Aratus, *Phaenomena* 27, 91; Ovid, *Metamorphoses* 2.40; *Fasti* 2.156.

Ps-Eratosthenes is the sole authority for Arcas marrying his mother. According to the Scholiast on Germanicus BP 64.15, Arcas attempted to use force against Callisto; Hyginus says that Arcas was hunting when he saw Callisto and pursued her into the sacred precinct.

According to Apollodorus, *The Library* 3.8.2, and Pausanias, *Guide to Greece* 8.3.6, when Callisto was shot by Artemis, Arcas was rescued by Zeus and brought by Zeus or Hermes to Maia to be nursed. Scholiast on Theocritus, *Idylls* 1.123 makes no mention of Arcas, but records that Hermes watched over Callisto on Mt. Lycaeum after she was changed into a bear.

2. As to the cult of Zeus Lycaeus, see Cook, *Zeus* 1:67–99; Pausanias, *Guide to Greece* 8.2.1. As to the sacred precinct of Zeus Lycaeus, see p. 200. Evidence for human sacrifice to Zeus Lycaeus may be found in Pausanias, *Guide to Greece* 8.38.7; Plato, *Republic* VIII (565D); [Plato], *Minos* (315C). The validity of the above evidence is questioned by R. P. Eckels, *Greek Wolf-Lore*, 53–55.

On Lycaon, see W. Drexler, "Lycaon," in Roscher, 2,2:2168–73; Pausanias, *Guide to Greece* 8.2.3; Ovid, *Metamorphoses* 1.208–11; Servius on Vergil, *Aeneid* 1.731, *Eclogues* 6.41; Apollodorus, *The Library* 3.8.1; Hyginus, *Fables* 176.

The changing of Lycaon into a wolf is reflected in the later belief that the transformation of a man into a wolf occurred regularly at the sacrifice to Zeus Lycaeus. The belief was that the man who tasted the human entrail among the sacrificial meats was turned into a wolf for a period of nine years. If he tasted of no human flesh during that time, he regained his human form at the end of the nine-year period, otherwise he remained a wolf for the rest of his life. On the significance of Lycaon's transformation into a wolf, see J. Fontenrose, "Philemon, Lot, and Lycaon," *University of California Publications in Classical Philology* 13 (1950): 93–120. As to the ritual of the sacrifice to Zeus Lycaeus, see the discussion in Cook, *Zeus* 1:67. See also Pausanias, *Guide to Greece* 8.2.38; Pliny the Elder, *Natural History* 8.81; St. Augustine, *The City of God* 18.17.

The city of Trapezus was situated in southwestern Arcadia near Mount Lycaeum. Concerning the origin of the name there are two traditions, both of them connected with members of Lycaon's family. Ps-Eratosthenes follows the more common tradition in deriving the name of the city from the table (*trapeza*) which Zeus overturned in Lycaon's house (see Apollodorus, *The Library* 3.8.1). For the other tradition, according to which the city was named after its founder, Trapezeus, a son of Lycaon, Pausanias is the sole authority.

On the location of the ancient Trapezus, already in ruins by the second century C.E., see Pausanias, *Guide to Greece* 8.29.1; E. Curtius, *Peloponnesus*, 1:304–6; W. M. Leake, *Travels in the Morea*, 2:pl. 2 (map).

3. For a discussion of Ps-Eratosthenes's use of Hesiod, see Franz, "De Callistus Fabula," *Leipziger Studien* 12 (1890): 306; W. Sale, "The Story of Callisto in Hesiod," *Rheinisches Museum* 105 (1962): 122–41.

4. Homer, *Iliad* 18.486, *Odyssey*, 5.272. Homer refers to Ursa Major as "The Wagon," see *Iliad* 18.487, *Odyssey* 5.273. Hesiod, writing about one hundred years after Homer, refers to Arcturus ("Bear-guard"), but is probably referring to the star rather than to the constellation; see *Works and Days* 610.

Philomelus is an obscure figure, mentioned only by Hyginus among classical authors.

5. As to the history of this constellation, see Boll and Gundel, 886–92; Allen, *Star Names*, 92–106.

6. See pp. 69–70 and 206.

7. H. W. Parke, *Festivals of the Athenians*, 118–19; Farnell, *Cults* 5:194–95.

8. Hesiod, *Works and Days* 582–88.

9. Compare English "dog-days" and French "canicule." See pp. 67 and 69–70.

Cancer

1. Although both Ps-Eratosthenes and Hyginus cite Panyassis as their source for the origin of the constellation Cancer, it is not clear whether Panyassis included the changing of the crab into a constellation in his account. Hyginus cites a variation of the story about Asini, substituting the sound of Triton's shell for the braying of asses—with similar result. Lerna, the site of Heracles's second labor, was a swampy area on the western coast of the Argolic Gulf. On Heracles, see pp. 101–3, 115–18, and 175–78.

2. The crab sent by Hera is also mentioned by Apollodorus, *The Library* 2.5.2. For other rivals of Hera and her treatment of them, see the stories of Callisto (pp. 197–99), Leto (Apollodorus, *The Library* 1.4.1), Io (Apollodorus, *The Library* 2.1.3), Semele (Apollodorus, *The Library* 3.4.3).

3. On the connection of Dionysus with donkeys, see W. Otto, *Dionysos*, chapter 14. None of the epithets of Dionysus reflects his connection with donkeys. One vase painting shows Dionysus seated in a ship whose bow is shaped like the head of a donkey (see J. Beazley, *Attic Black-Figure Vase Painting*, 369, no. 1001). On the François-vase, Hephaestus is shown riding a donkey (see A. Minto, *Il vaso François*). On the connection between Hephaestus and Dionysus, see L. Malten, "Hephaistos," in *Realenzyklopädie*, 8:311. On satyrs, see A. Hartmann, "Silenos und Satyros," in *Realenzyklopädie*, 2A,3:35–53.

See also Boll and Gundel, 951–54; Allen, *Star Names*, 107–14.

Canis Major

1. On Odysseus's dog see Homer, *Odyssey* 17.291–327. For the story

of Hecuba, see Euripides, *Hecuba*. Monstrous dogs, such as the three-headed Cerberus and the hell-hounds of Hecate, are associated with the Underworld. On Cerberus, see Apollodorus, *The Library* 2.113. The hounds of Hecate are described by Apollonius Rhodius, *The Voyage of Argo* 3.1211ff.

2. Procris's dog was originally fashioned out of bronze by Hephaestus, who brought it to life and presented it to Zeus. See Apollodorus, *The Library* 3.15.1. Procris herself was, by most accounts, the daughter of Erechtheus (see p. 51). On Minos's illness and subsequent cure by Procris, see Antoninus Liberalis, *Metamorphoses* 41. The illness was caused by Pasiphae, who resented Minos's intimacies with other women and contrived a charm as a result of which any woman who had intercourse with him died.

On Orion, see pp. 147–50. On the story of Icarius's dog, see pp. 57–58.

3. Marvelous weapons and animals are common themes in folk-tale: see Thompson, *Motif-Index*, D1080: "Magic weapons"; B100: "Magic animals."

4. See Homer, *Iliad* 22.29. Hipparchus, 2.1.18 and Ptolemy, *Almagest* 7.6, report that both the constellation and its brightest star were known as *Cyon* ("Dog"). See also Aratus, *Phaenomena* 326. Variant readings in the manuscripts render the Greek and Latin texts confusing and sometimes contradictory. Some manuscripts distinguish between a star on the head called Isis and one on the tongue called Sirius or Canis/Cyon; others reverse those references or speak of the two as one star. Ptolemy differentiates between "the head" (μ CMa) and "the mouth, the one called Cyon" (α CMa).

5. There are numerous references to Sirius as the heat-bringer in Greek and Latin literature: see, e.g., Hesiod, *Works and Days* 587; Vergil, *Georgics* 4.425–28.

6. See Boll and Gundel, 995–1002; Allen, *Star Names*, 117–31. On the significance of Sirius to the Persians, see Plutarch, *On Isis* 47.

Canis Minor

1. On the name Procyon, see Scholiast on Aratus, *Phaenomena* 450. Horace, *Odes* 3.29.18, apparently confuses Procyon with Sirius. Gundel, "Procyon," in *Realenzyklopädie*, 23, 1:622 suggests that Procyon is to be considered a doublet of Canis Major. Hyginus, *Poetic Astronomy* 2.4 may erroneously equate the Latin name Canicula with Procyon. Canicula appears to have been used by Latin authors with reference to Canis Major. See Boll and Gundel, 1003.

On Orion and Canis Major, see pp. 147–50 and 65–67. On the Hare, see pp. 129–31. The "other wild animals" mentioned as being beside Procyon are, possibly, Cancer (see pp. 61–63), Leo (see pp. 125–26), Ursa Major (see pp. 197–99). Lepus, which is said in the present story to be close to Procyon, is in reality closer to Canis Major. It is entirely possible that Ps-Eratosthenes has confused the two constellations of Canis Major and Procyon, as all his references to Procyon seem to be more suited to Canis Major. Scholiast on Aratus, *Phaenomena* 450 and Scholiast on Germanicus BP 102.1 refer, more correctly, to the proximity of Ursa Major to Procyon.

2. On the name Cyon, see p. 65. On the Euphratean and Egyptian background of Procyon, see Boll and Gundel, 1003–4; Allen, *Star Names*, 131–35.

Capricorn

1. Aegipan is said to be the son of Zeus and Aega, the wife of Pan (Hyginus, *Poetic Astronomy* 2.13), or of Zeus and Boetis (Hyginus, *Fables* 155). See also W. H. Roscher, "Die Sagen von der Geburt des Pan," *Philologus* 53 (1894): 362–77. Pan was believed to be the father of Iynx by Echo or Peitho.

The Constellations refers to Aegipan as the ancestor of Capricorn. This is not supported by Greek myths, as the genealogies of Pan and Aegipan do not coincide. The literary references to Aegipan are late (Apollodorus, *The Library* 1.4.11; Hyginus, *Fables* 155, *Poetic Astronomy* 2.13, 28). The only myth connected with Aegipan is his stealing, in company with Hermes, of the sinews of Zeus which had been hidden by Typhon during the Gigantomachia: see Apollodorus, *The Library* 1.4.11; Nonnus, *Dionysiaca* 1.481.

Descriptions of Pan are given by *Homeric Hymns* 19; Lucian, *Dialogues of the Gods* 22. Note also the epithets *aigokeros* ("goat-horn"), *aigiknamos* ("goat-limbed"), *dikeros* ("two-horned"), *tragopous* ("goat-foot") applied to Pan: see Bruchmann, *Epitheta Deorum*, 185–89.

For the references to Pan in Classical literature and art, see K. Wernicke, "Pan," in Roscher, *Lexikon* 3,1:1406–71.

2. Aegipan is never described in Greek literature, but his appearance may be inferred from the description of Capricorn in *The Constellations*.

Pan, although he is regularly a patron of fishermen and coast-dwellers, is never represented in Greek literature or art as having a fish's tail: see *Palatine Anthology* 10.10; Theocritus, *Idylls* 5.14.

As to the Babylonian constellation-figure represented as half-goat, half-fish, see Boll and Gundel, 972.

Although most artistic representations of this constellation show the Babylonian half-goat, half-fish figure, *The Constellations* describes a half-man, half-goat/fish creature. Aratus, *Phaenomena* 284 and Ptolemy, *Almagest* 8, mention the fish's tail of this figure.

3. The name Capricorn is, however, an epithet of Pan: see Hyginus, *Fables* 196. As to the infancy of Zeus on Mt. Ida, see pp. 50 and 52. There was a cult of Pan on Crete: see "Pan," in Roscher, *Lexikon*, 1372.

4. See "Titan," in *Dictionary of Greek and Roman Biography and Mythology* (ed. W. Smith), 3:1156–57. The Titanomachia is regularly set in Thessaly: see, e.g., Hesiod, *Theogony* 629. The Gigantomachia was usually located at Phlegrai, but also at various other places: see Apollodorus, *The Library* 1.6.1–3, and vol. 1, p. 43, note 3 in the Loeb Classical Library edition.

5. As to the shell-trumpet, which inspired "panic" fear, see Scholiast on Aratus, *Phaenomena* 284; Scholiast on Germanicus BP 87.3, G 155.19. As to the trumpet of Triton, see Ovid, *Metamorphoses* 1.333–42.

6. The myths connected with this constellation are discussed by W. H. Roscher, "Die Elemente des astronomischen Mythus vom Aigokeros," *Fleckeis. Jahrbücher* 151 (1895): 333–42, and Boll and Gundel, 973. The story of Pan's metamorphosis into a goat/fish is also related by Hyginus, *Fables* 196; Antoninus Liberalis, *Metamorphoses* 28; Ovid, *Metamorphoses* 5.325. The Nereid in place of Capricorn is found in Teucrus and Antiochus (in Boll, *Sphaera*, 277–78).

As to the history of this constellation, see Allen, *Star Names*, 135–42; Boll and Gundel, 971–74.

Cassiopeia

1. On Medusa's boast and subsequent punishment, see Apollodorus, *The Library* 2.4.3. On Agamemnon's boast, see Apollodorus, *Epitome* 3.21. Cassiopeia's challenging of the Nereids is recounted by Apollodorus, *The Library* 2.4.3. Cassiopeia appears to be in an upside-down position during half her revolution about the north pole.

2. See also pp. 27–28, 83–84, 85, 157–58. On the history of this constellation, see Allen, *Star Names*, 142–48.

3. Apollodorus, *The Library* 1.2.7, lists forty-five Nereids by name. For other lists, see Homer, *Iliad* 18.38–49 (lists thirty-one); Hesiod, *Theogony* 240–64 (lists fifty-one); Hyginus, *Fables* preface (lists thirty-two). On the wooing of Amphitrite by Poseidon, in connection with the constellation Delphinus, see pp. 97–100.

The name and something of the character of the Nereids have survived in the Modern Greek *neraides*. The latter are thought of as mermaids or water-nymphs, but the name is also used of the nymph-spirits of forests, valleys, and ridges. For a discussion of the *neraides*, see J. C. Lawson, *Modern Greek Folklore and Ancient Greek Religion*, 132–34.

Centaurus

1. As to the wisdom and justice of Chiron, see Homer, *Iliad* 11.832; Xenophon, *On Hunting* 1.1; Ovid, *Fasti* 5.384, 413; Pindar, *Pythian Odes* 9.64; Horace, *Epodes* 13.11. Chiron's abode was said to be on Mt.

Pelion, in a cave overlooking the Pagasitic Gulf, see Pindar, *Pythian Odes* 9.30. Chiron appears on coins of Magnesia: see B. V. Head, *Historia Numorum*, 300–301. Apollodorus, *The Library* 2.5.4, says that Chiron dwelt at Malea in the Peloponnesus after being driven from Mt. Pelion by the Lapithae. As to the pupils of Chiron, see the list in Xenophon, *On Hunting* 1.1. Chiron's skills in healing are referred to by Homer, *Iliad* 4.219, 11.831. Pliny, *Natural History* 7.196, says that Chiron was the founder of herb-medicine. The Scholiast on Homer, *Iliad* 4.219, refers to Chiron's musicianship. Chiron was also a prophet: at the wedding of Peleus and Thetis, he foretold the greatness of Achilles (Euripides, *Iphigenia at Aulis* 1064). Chiron's friendliness to man is mentioned by Pindar, *Pythian Odes* 3.5; Scholiast on Apollonius of Rhodes, *Voyage of Argo* 1.554; Homer, *Iliad* 16.143; Apollodorus, *The Library* 3.13.5.

2. See Apollodorus, *The Library* 1.2.4; Pliny, *Natural History* 7.197; Ovid, *Metamorphoses* 6.126. See also pp. 150–53.

3. On Chiron's immortality, see Aeschylus, *Prometheus Bound* 1027; Sophocles, *Maidens of Trachis* 715; Lucian, *Dialogues of the Dead* 26. The story of Chiron's death is related by Apollodorus, *The Library* 2.5.4; Ovid, *Fasti* 5.398; Scholiast on Germanicus BP 99.17, G 178.14.

4. See p. 185.

5. The name Hippocrator is given to Centaurus by Hermes Trismegistus, 252–253G. As to the history of the constellation, see Boll and Gundel, 1012–14; Allen, *Star Names*, 148–155.

6. As to Lupus, see Boll and Gundel, 1014–16; Allen, *Star Names*, 278–79. Lupus is comprised of ten stars according to *The Constellations* and to Hipparchus, nineteen according to Ptolemy. As to the Altar, see pp. 37–38.

7. As to the thyrsus, see F. v. Lorentz, "Thyrsos," in *Realenzyklopädie*, 6A:747–52.

Cepheus

1. G. P. Goold, "Perseus and Andromeda: A Myth from the Skies," *Proceedings of the African Classical Association* 2 (1959): 10–15, argues for

the astronomical origin of the Perseus-Andromeda myth.

2. Cepheus generally appears in classical literature as the king of Ethiopia or Phoenicia, but Media and Iope are sometimes mentioned as being under his rule. See Tacitus, *Histories* 5.2; Pliny, *Natural History* 6.183. Greek authors are not in complete agreement as to the location of Ethiopia. Aeschylus connects the Ethiopians with India (*Suppliants* 284), but beginning with Herodotus, the name Ethiopia is applied to the area south of Egypt. On Cassiopeia, see pp. 75–76; on Andromeda, see pp. 27–28; on Perseus, see 157–58; on Cetus, see pp. 85–86. The fullest account of the story of Cassiopeia's arrogance and the resulting exposure of Andromeda to the sea monster is in Apollodorus, *The Library* 2.4.3.

3. On the origin of this constellation, see Allen, *Star Names*, 155–159. Achilles Tatius, *Isagoga* 75M, (third century C.E.) states that the constellation of Cepheus was not known to the Chaldeans or the Egyptians.

Cetus

1. For a general description of this constellation figure, see Aratus, *Phaenomena* 353–66. The name Cetus is from the Greek word meaning "sea creature" and is used in Greek of whales and other large sea animals, not necessarily with the connotation of "monster."

Ceto appears in Greek mythology as a sea-nymph, the daughter of Pontus ("Ocean") and Gaea ("Earth"). See Apollodorus, *The Library* 1.2.6; Hesiod, *Theogony* 238.

See also pp. 27–28, 75–76, 83–84, 157–58.

2. Brown, *Primitive Constellations*, 1:88–91, 189, 2:55; Allen, *Star Names*, 160–64.

Corona Borealis

1. Theseus received the crown as a gift from Thetis, who had received it as a wedding gift from Aphrodite, according to Bacchylides, 17.112.

2. The earliest literary reference to Ariadne's crown is in Pherecydes

(see Scholiast on Homer, *Odyssey* 11.321). According to Pherecydes, the golden crown was presented to Ariadne by Dionysus on Naxos; it is doubtful that Pherecydes associated Ariadne's crown with the constellation. Pherecydes is followed by Apollonius Rhodius, *The Voyage of Argo* 3.1001, who mentions that the crown was turned into a constellation by the gods. Servius on Vergil, *Georgics* 1.222, relates that the crown was a gift from Dionysus and was placed in the heavens by that god. Aratus, *Phaenomena* 71, says that the crown was placed in the sky by Dionysus in his grief over the death of Ariadne. The Scholiast on Aratus, *Phaenomena* 71, says that the constellation was the ivy wreath worn by Dionysus. As to depictions of the crown of Ariadne on ancient vase-paintings, see H. Steuding, "Theseus," in Roscher, *Lexikon*, 5:694, 696.

3. See Farnell, *Cults*, 2.632–35. Ariadne was the mother of Tauropolis according to the Scholiast on Apollonius Rhodius, *The Voyage of Argo* 3.997. On the worship of Ariadne at Athens, see Plutarch, *Theseus* 22.

4. See Scholiast on Theocritus, *Idylls* 2.45; Scholiast on Apollonius Rhodius, *The Voyage of Argo* 3.997; Diodorus Siculus, 4.61.5, 5.51.3.

5. Ariadne was slain by Artemis according to Homer, *Odyssey* 11.321. The more common story of her abandonment by Theseus and subsequent marriage to Dionysus is alluded to by Hesiod, *Theogony* 947; Pausanias, *Guide to Greece* 1.20.3; Hyginus, *Fables* 43; Plutarch, *Theseus* 20. The grave of Ariadne was shown on Naxos, on Cyprus, and at Argos. For the cult of Ariadne on Naxos, see H. Stoll, "Ariadne," in Roscher, *Lexikon*, 1,1:540–46. On the grave of Ariadne at Argos see Plutarch, *Theseus* 20; Pausanias, *Guide to Greece* 2.23.8.

Ancient references point to the use of both crowns and wreaths in the marriage rite. See W. K. Lacey, *The Family in Ancient Greece*, plates 24, 26.

6. See Apollodorus, *The Library* 3.5.3. On the descent of Orpheus to the Underworld, and his connection with Dionysus, see W. K. C. Guthrie, *Orpheus and Greek Religion*, 29–32, 53.

7. See Boll and Gundel, 892–96; Allen, *Star Names*, 174–79. Another "crown" is located by Ptolemy in the vicinity of Sagittarius. The

Scholiast on Aratus, *Phaenomena* 400, refers to that group of stars as the Southern Crown [Corona Australis], or the wheel of Ixion.

As to Ariadne's Lock, see pp. 125–28.

Cygnus

1. On the raising of Helen by Leda see Apollodorus, *The Library* 2.10.7; Pausanias, *Guide to Greece* 1.33.7; Scholiast on Callimachus, *Diana* 232.

The birth of Helen from an egg laid by Leda is recounted by Lucian, *Dialogues of the Gods* 20.14; Scholiast on Homer, *Odyssey* 11.298; Hyginus, *Fables* 77; Apollodorus, *The Library* 3.10.7. The egg supposed to have been laid by Leda was seen by Pausanias at Sparta (Pausanias, *Guide to Greece* 3.16.1) in the second century C.E. Plutarch, *Symposium* 637B recounts that the egg fell from heaven.

According to Homer, *Odyssey* 4.227, 569, Helen was the daughter of Zeus. Hesiod, Fragment 92Rz says she was the daughter of Zeus and one of the daughters of Oceanus. For other accounts of Helen's birth, as well as the parentage of her brothers and sister, see R. Engelmann, "Helena," Roscher, *Lexikon*, 1.2:1932. Helen's brothers, Castor and Pollux, were transformed into the constellation Gemini (see pp. 111–13).

2. The concept of righteous indignation (*aisa*), whether of men or of gods, is present in Homer but is not personified.

3. On the cult of Nemesis at Rhamnus, see Farnell, *Cults*, 2.487–94; Cook, *Zeus*, 1.265–85; Pausanias, *Guide to Greece* 1.33.2–8. Rhamnus was situated on the east coast of Attica, to the north of Tricorythus. See Strabo, *Geography* 9.1.22.

Farnell believes that the early goddess of Rhamnus was a divinity of life and death similar to Artemis and Aphrodite, but he is uncertain of the meaning of the epithet Nemesis, as it cannot have meant "divine retribution" in connection with the earlier cult. He suggests that the meaning of Nemesis might be sought in Gk. *nemo* ("to deal out, dispense"), since a goddess of birth could be associated with the dispensing of a particular lot in life.

The cult at Smyrna was to two goddesses named Nemesis, who were daughters of Night (see Hesiod, *Theogony* 223; Pausanias, *Guide to Greece* 7.5.2). These goddesses, who were probably only one goddess, were nature divinities connected with vegetation.

As to the identification of Nemesis with Diana Nemorensis, see Cook, *Zeus*, 1:273–75. Cook supports his argument with the contention that the cult of abstractions is late, while that of Nemesis is early, although he admits that the Rhamnusian Nemesis is not an abstraction, but an anthropomorphized figure.

4. On the history of this constellation, see Boll and Gundel, 906–8; Allen, *Star Names*, 192–98. The constellation is called simply the "Bird" by Aratus, *Phaenomena* 275 and Ptolemy, *Almagest* 7.5.

5. The first myth is attested by Scholiast on Aratus, *Phaenomena* 273, 275. As to the constellation of the Lyre, see pp. 133–39. The story of Cygnus, who was changed into a swan, is told by Ovid, *Metamorphoses* 2.377; Vergil, *Aeneid* 10.189. The changing of Cygnus into a constellation is alluded to by all the above except Ovid.

Delphinus

1. The dolphin of classical literature and art is the Delphinus delphis (on which, see Keller, *Thiere des classischen Altertums*, 211–35; *Antike Tierwelt*, 1:408–9; Stebbins, *The Dolphin in the Literature and Art of Greece and Rome*, 1–8) and not the Coryphaena hippurus, known in English as the Common Dolphin. The appearance of Delphinus delphis is described by Stebbins, *The Dolphin*, 4: "Delphinus delphis does not grow over ten feet in length. The salient features of its head are the round brain case with the fatty cushion in front of the blowhole and the long beak. At the point where these meet there is a V-shaped groove. The mouth or beak . . . is long and furnished with forty to sixty pairs of teeth in either jaw."

Accounts of the dolphin's friendship and service to man include the myths of Coeranus, Melicertes-Palaemon, Eualus, Arion, and numerous folktales (Stebbins, *The Dolphin*, 62–70, 73–77).

2. Stebbins, *The Dolphin*, 97–129. The story of Dionysus and the

Tyrrhenian pirates was also well known.

3. On the connection of Poseidon with dolphins, see Stebbins, *The Dolphin*, 84–86 (literary references). Poseidon is represented with the dolphin on coins of Boeae, Gythion, Caphyae, Sybaris, Poseidonia, Corinth, Tenos, Caria, Galatia, Cilicia, Alexandria, etc. See Farnell, *Cults*, 4:96–97.

4. Amphitrite is mentioned as the consort of Poseidon as early as Hesiod, *Theogony* 930. *The Constellations* provides the earliest account of her wooing by Poseidon. Amphitrite was never worshipped alone, but always together with Poseidon. There is evidence of this double worship at several of the better known centers of the Poseidon cult: Tenos, Syros, Myconos, the Isthmus of Corinth, Lesbos, Amyclae. See Farnell, *Cults*, 4:1–60.

Little is known about Artemidorus aside from the fact that he was a pupil of Callimachus.

5. See *Scholia to Aratus* 318. The leading of the Cretans to Delphi by Apollo in the shape of a dolphin is widely attested: see Stebbins, *The Dolphin*, 77–80. There is evidence of a cult of Apollo Delphinius at Chalcis, Athens, Aegina, Sparta, Thera, Chios, Crete, Massilia, Miletus, Olbia. See Farnell, *Cults*, 4:145–48, and notes.

6. As to the dolphin's love of music, see Aelian, *The Nature of Animals* 2.6, 11.12, 12.45; [Arion], *Hymn to Poseidon* 8; Pliny the Elder, *Natural History* 9.24; see also Martianus Capella, *The Marriage of Philology and Mercury* 9.927; Archias in *Palatine Anthology*, 7.214; Euripides, *Electra* 435–36; Plutarch, *The Banquet of the Seven Wise Men* 19 (162E).

The evidence from art tends to derive the dolphin's affinity for music from its being one of the creatures sacred to Apollo, e.g., an Attic red-figured vase shows Apollo Delphinius holding a lyre and flanked by dolphins: see Stebbins, *The Dolphin*, 104.

7. On the history of this constellation, see Allen, *Star Names*, 198–201; Boll and Gundel, 926–27.

Draco

1. Hesiod, *Theogony* 215, 275, 518. The serpent is sometimes called

Dracon, but usually is not given a name. (The Greek word *drakon* means "snake" and is interchangeable with the more common word *ophis*; see H. G. Liddell and R. Scott, *Greek-English Lexicon*, 9th ed., 448). Apollonius Rhodius, *The Voyage of Argo* 4.1396 calls the serpent Ladon. The existence of a river of this name in the northwest Peloponnesus has caused some scholars to locate the garden of the Hesperides in Arcadia (see K. Seeliger, "Hesperiden," in Roscher, *Lexikon* 1,2:2594–2603). The garden is located in the West by most ancient sources, as the mention of the Atlas Mountains would suggest. Hesiod says the garden of the Hesperides was "beyond Ocean."

2. Although cited as a source by Ps-Eratosthenes and subsequent authors, including Hyginus and the scholiasts on Germanicus (BP 60.15, G 118.18), Pherecydes does not identify the serpent with the constellation. In Fragment 16J, Pherecydes recounts how Heracles sent Atlas to fetch the apples and did not himself meet the dragon. See also Apollodorus, *The Library* 2.5.11. On Heracles, see pp. 115–18 and 175–78.

3. Scholiast on Aratus, *Phaenomena* 45, 46 (Python, or dragon of Cadmus, or snake of Zeus); Hyginus; Scholiast on Germanicus BP 60.15, G 118.18 (dragon of Athena); Scholiast on Homer, *Odyssey* 5.272 (snake of Zeus).

4. Thompson, *Motif-Index of Folk-Literature*, B11.6.2: "Dragon guards treasure"; F480.2: "Serpent as house spirit."

5. On the history of this constellation, see Allen, *Star Names*, 202–12; Boll and Gundel, 821–24.

Eridanus

1. See Aratus, *Phaenomena* 358–60; Scholiast on Germanicus BP 98.6; G 174.20; Nonnus, *Dionysiaca* 38.429; Hermes Trismegistus, 198G.

2. On the Eridanus River, see Hesiod, *Theogony* 33; Herodotus, *The Histories* 3.115; Strabo, *Geography* 5.1.9; Pausanias, *Guide to Greece* 1.4.1, 19.5; 5.12.7, 14.3; 8.25.13; Vergil, *Georgics* 1.482, 4.371; *Aeneid* 6.659; Hyginus, *Fables* 154; Pliny the Elder, *Natural History* 3.117;

37.31.

As to the association of the Eridanus with amber, see Ovid, *Metamorphoses* 1.750–2.366; E. H. Warmington, *Greek Geography*, xxx; K. Dilthey, *De Electro et Eridano*. See also Pliny the Elder, *Natural History* 37.30–53. As to the three main routes used for the transport of amber in ancient times, see J. M. de Navarro, "Prehistoric Routes between Northern Europe and Italy defined by the Amber Trade," *Geographical Journal* 66 (1925): 481–504.

3. As to the story of Phaethon, see Scholiast on Homer, *Odyssey* 17.208; Ovid, *Metamorphoses* 1.750–2.366; Lucian, *Dialogues of the Gods* 25; Hyginus, *Fables* 152, 154.

4. As to the star Canopus, see Allen, *Star Names*, 67–72. On the visibility of Canopus from Greece, see Scholiast on Aratus, *Phaenomena* 351.

As to Canopus, the helmsman of Menelaus, see Conon, 8; Strabo, *Geography* 17.801; Plutarch, *On the Deceitfulness of Herodotus* 12 (857B), *On Isis* 22 (says that Canopus was the helmsman of Osiris and was placed among the stars together with his ship [Argo], on which see pp. 39–42).

5. As to the history of this constellation, see Allen, *Star Names*, 215–20; Boll and Gundel, 989–93. As to the Euphratean "Sea," see Manilius, *Astronomy* 1.440; Teucrus, Antiochus, Valens (in Boll, *Sphaera*, 134–36). Brown, *Eridanus: River and Constellation*, contends that the River of the Greeks actually represents the Euphrates River (the Nile and the Euphrates were believed by the ancients to be different parts of the same river). As to the artistic representations of the River, see Thiele, *Antike Himmelsbilder*, 39–40; Boll and Gundel, 990–92.

6. There is no mention of the number of stars in the River in the extant writings of Hipparchus.

Galaxy

1. As to the heavenly circles, see Achilles Tatius in Maass, *Commentariorum in Aratum Reliquiae*, 51.

240 NOTES TO PAGES 110–112

2. See Diodorus Siculus, *History* 4.9.6; Pausanias, *Guide to Greece* 9.25.2 (these two authors do not relate the story as an aetion for the Milky Way); Achilles Tatius (in Maass, *Commentariorum*, 55); Scholiast on Germanicus BP 186.25. A late allusion to the myth (Nonnus, *Dionysiaca* 35.308–11) makes Dionysus the recipient of the divine milk instead of Heracles.

As to the myth of Phaethon in connection with the Milky Way, see Manilius, *Astronomy* 1.735–42; Aristotle, *Meteorologica* 345a.

As to the changing of the Sun's path in consequence of the sacrilege of Thyestes, who slew his son and placed the cooked parts before the gods, see Achilles Tatius (in Maass, *Commentariorum*, 55).

3. Pindar, *Hymns* 30; Ovid, *Metamorphoses* 1.168–76; Martianus Capella, 22.208; Lucian, *In Praise of Demosthenes* 50; Quintus Smyrnaeus, *The Fall of Troy* 14.223; Stobaeus, *Anthology* 1.574, 906; Scholiast on Germanicus BP 187.14.

4. Boll and Gundel, 1022, 1028. A Modern Greek myth explaining the Milky Way as the trail left by a grain-thief as he fled in haste, is said to derive from an Ancient Near Eastern story which may have been known to Eratosthenes. See Boll and Gundel, 1026.

Artistic representations of the Milky Way are not common. See Thiele, *Antike Himmelsbilder*, 42.

Gemini

1. As to the parentage of the Dioscuri, see Homer, *Odyssey* 11.298–304; *Homeric Hymns* 17.2; Pindar, *Olympian Odes* 3.61; Apollodorus, *The Library* 3.11.2; Lucian, *Dialogues of the Gods* 26. According to Homer, Castor and Polydeuces were both mortal; according to the *Homeric Hymn*, they were both immortal. In the remaining sources one brother was immortal, usually Polydeuces, and the other was mortal. See also pp. 93–94.

On the Dioscuri as Laconian heroes, see Farnell, *Greek Hero Cults*, 175–228. There is evidence for cults of the Dioscuri in the Peloponnesus, Attica, Northern and Northwestern Greece, Delphi, Corcyra, Cephallenia, Thessaly, and Magna Graecia.

2. On twins in general, see J. R. Harris, *The Cult of the Heavenly Twins*, 4–62. The name Dioscuri means "children of Zeus." On the Dioscuri, see Cook, *Zeus*, 1:760–75, 2:422–40. Twins are generally connected in cult with their mother. The Dioscuri are often connected with their mother, Leda, but also with their sister, Helen. See F. Chapouthier, *Les Dioscures au service d'une déesse*. On other sets of twins in Greek mythology, see Cook, *Zeus* 2:444; C. Kerenyi, *The Heroes of the Greeks*, 34–39, 105–12. The Great Twins in Babylonian art are represented head to head or feet to feet as well as side by side (see Brown, *Primitive Constellations*, 2.42).

3. Twins are often thought of as horsemen, coming especially to the aid of those in danger at sea, and the Dioscuri are no exception. See Pliny the Elder, *Natural History* 2.101. Euripides, *Orestes* 1637, says that Helen is a savior of mariners. See also K. Jaisle, *Die Dioskuren als Retter zur See bei Griechen und Romern*, for a study of the relevant passages in classical literature. According to Livy, *History of Rome* 2.19, the Dioscuri came to the aid of the Roman legions at the battle of Lake Regillus *c.* 496 B.C.E.

4. On Gemini as Amphion and Zethus, see Scholiast on Germanicus BP 68.7; as Heracles and Apollo, see Manilius, *Astronomy* 4.755; as Heracles and Theseus, see Scholiast on Germanicus BP 69; Hyginus, *Fables* 55; as the Great Gods of Samothrace, see Scholiast on Euripides, *Orestes* 1637, as Phaon and Satyrus, see Hermes Trismegistus in W. Gundel, *Neue Astrologische Texte des Hermes Trismegistos*, 263–64. As to the history of this constellation, see Boll and Gundel, 946–51; Allen, *Star Names*, 222–37.

Hercules

1. The winning of the golden apples was either the eleventh or the twelfth labor of Heracles. See pp. 101–3, 177. On the labors in general see Apollodorus, *The Library* 2.5.1–12.

2. See Aratus, *Phaenomena* 63–66. On Theseus, see pp. 87–89. Thamyris was a Thracian bard who challenged the Muses in song; he was blinded by them in punishment (Homer, *Iliad* 2.594 and Scholiast).

On Orpheus, see pp. 133–35. Ixion attempted to force himself on Hera. His punishment in the Underworld was to be bound to a wheel that turned eternally (Pindar, *Pythian Odes* 2.21 and Scholiast). When the figure is identified with Ixion, the constellation Corona Borealis (see pp. 87–91) is considered to be Ixion's wheel. On Prometheus, see pp. 175–78. Greek myth assigns to Atlas the task of holding up the sky (Hesiod, *Theogony* 517–20). Tantalus purportedly served the flesh of his son to the gods and was punished in the Underworld by being eternally "tantalized," that is, hungry and thirsty, and never satisfied (Homer, *Odyssey* 11.582–92). The identification with Ceteus, a son of Lycaon, connects this constellation with Bootes and the two Bears.

Hydra, Crater, Corvus

1. See Ovid, *Fasti* 2.243; Scholiast on Aratus, *Phaenomena* 449; Scholiast on Germanicus BP 100.12, G 180.15; Aelian, *On the Characteristics of Animals* 1:47, 12.4; Catullus, 66.57. See also Thompson, *Glossary of Greek Birds*, 159–64. The hawk is also closely connected with Apollo, as are several other birds, and is often referred to as that god's messenger.

As to the myth of Coronis, see Apollodorus, *The Library* 3.10.3; Pausanias, *Guide to Greece* 2.26.6; Hyginus, *Fables* 202; Ovid, *Metamorphoses* 2.534; Antoninus Liberalis, *Metamorphoses* 20.7.

2. As to the Perseus-Andromeda group, see pp. 27–28, 75–76, 83–84, 85, 157–58.

The story of the crow and the figs is a conglomeration of folk-tale motifs: see Thompson, *Motif-Index*, B291.1.2: "Crow as messenger"; F420.4.9: "Water-supply controlled by water-spirit"; K401.2.1: "Crow tricks snake"; A2435.4.3: "Why raven suffers thirst." For other fables concerning the crow, see Thompson, *Glossary of Greek Birds*, 95.

See Aelian, *Characteristics of Animals* 1.47; Aristotle, *Fragment* 343R3. The treatise of Archelaus is not extant.

3. For Corvus as the bird touching the head of Talas, see Boll, *Sphaera*, 279. The Hydra is identified with the hydra slain by Heracles at Lerna (Scholiast on Aratus, *Phaenomena* 443), the serpent guarding

the apples of the Hesperides (Manilius, *Astronomy* 5.16), with the Nile River (Scholiast on Aratus, *Phaenomena* 443). As to Icarius, see pp. 56–58. Otus and Ephialtes were twin sons of Poseidon, who imprisoned the god Ares in a bronze vessel because he had slain Adonis (see Scholiast on Homer, *Iliad* 5.385).

4. On the history of these three constellation-figures, see Allen, *Star Names*, 179–84, 246–50; Boll and Gundel, 1008–12.

The city of Eleusa was located at the entrance to the Hellespont, on the European side. A port called *Crateres Achaion* ("craters of the Achaeans") was located by [Scylax], *Periplus* 96 on the Chersonese.

Leo

1. On the slaying of the Nemean lion by Heracles, see Hesiod, *Theogony* 326; Hyginus, *Fables* 30; Apollodorus, *The Library* 2.5.1. The lion was said to have fallen from the sky or moon by Plutarch, *On the Face in the Moon* 24. The rearing of the lion by Hera may be an extension of her usual connection with the moon. See Cook, *Zeus* 1:453–57. On Heracles, see also pp. 61–62, 101–3, 115–18. Pisander of Rhodes (sixth century B.C.E.) wrote a *Heraclea* in two books, of which only fragments survive.

2. See pp. 33–34, where the explanation given for Aquila is that the eagle is the "king of birds."

According to the usual account, Heracles was forced to kill the Nemean lion with his hands when he found its hide invulnerable to his arrows. *The Constellations* is the only classical source to mention that Heracles strangled the lion with his bare hands because he sought after glory.

3. As to the history of this constellation, see Boll and Gundel, 954–59; Allen, *Star Names*, 252–63. On the connection of the lion with the sun, see Macrobius, *Saturnalia* 1.21.

4. As to Coma Berenices, see Pliny the Elder, *Natural History* 2.71. Callimachus's poem on the changing of Berenice's Lock into a constellation is preserved in the translation of Catullus (66).

5. Coma Berenices is comprised of three "unformed" stars accord-

ing to Ptolemy. According to Ps-Eratosthenes and Hyginus, the constellation consists of seven stars. These stars are said to comprise the mane of the Lion by Dorotheus (in *Catalogi Codicorum Astronomicorum* 6.94) and by Nonnus, *Dionysiaca* 2.655. Ptolemy says they represent an ivy-leaf. They are identified as a grape cluster by Cosmas Indicopleustes in Maass, *Analecta Eratosthenica*, 5, and as a spindle by Scholiast on Aratus, *Phaenomena* 144. Aratus himself does not name these stars.

Lepus

1. On Orion, see pp. 147–50; on Procyon, see pp. 69–70. The plague of hares on Astypalaea is mentioned by Athenaeus, *The Banquet* 9.400.

2. Scholiast on Aratus, *Phaenomena* 338; Scholiast on Germanicus G 170.17. On the epithets of Hermes, see C. Bruchmann, *Epitheta Deorum*, 111.

3. Aristotle, *The History of Animals* 542b, 579b–580a, 585a; *The Generation of Animals* 774a.

4. See Boll and Gundel, 993. The constellation Lepus rises when Corvus sets. Allen, *Star Names*, 265, sees a reflection of astronomical fact in Aelian, *On the Characteristics of Animals* 13.11, where the enmity of hares and crows is noted. On Corvus, see pp. 119–21.

Lyra

1. As to the shape and origin of the lyre, see *Homeric Hymns* 4.24–61; Apollodorus, *The Library* 3.10.2; Ovid, *Metamorphoses* 11.167; Hyginus, *Fables* 273. See also R. P. Winnington-Ingram, "Ancient Greek Music 1932–1957," *Lustrum* 3 (1958): 14, 5–57; I. Düring, "Greek Music," *Journal of World History* 3 (1956): 307; C. Sachs, *The Rise of Music in the Ancient World*, 218–20, 229; "Lyra," in Smith, *Dictionary of Greek and Roman Antiquities*. A list of vase paintings showing lyres has been compiled by O. Gombosi, *Tonarten und Stimmungen der antiken Musik*. The later cithara differed from the lyre in lacking the tortoise-shell base which served as a sounding chamber, and thus produced higher and less resonant tones than the lyre. The lyre was generally consid-

ered a more manly instrument than the cithara.

2. A collection of the testimonia relating to Orpheus will be found in O. Kern, *Orphicorum Fragmenta*, 1–79. For a discussion of Orpheus's connection with music, see I. M. Linforth, *The Arts of Orpheus*, 165–66. Two musicians of great repute and also of Thracian origin, Thamyris and Linus, were said to be pupils of Orpheus. Other pupils of Orpheus included Musaeus, Eumolpus, and Midas.

Although Orpheus was generally believed to have lived before Homer (eighth century B.C.E.), the image of Orpheus as a musician who could move all nature is not attested earlier than the fifth century B.C.E. Orpheus does not appear at all on black-figured vases, while on red-figured vases he is never shown surrounded by wild animals. See J. E. Harrison, *Prolegomena to the Study of Greek Religion*, 458. As to the magical powers of Orpheus's music, see Euripides, *Bacchae* 550, *Iphigenia at Aulis* 1213; Apollodorus, *The Library* 1.3.1; Pausanias, *Guide to Greece* 6.20.18; Horace, *Odes* 1.12.7; Ovid, *Metamorphoses* 10.86–105.

As to "Orphic" literature, the most complete collection of its remains can be found in Kern, *Orphicorum Fragmenta*, 80–344. The authenticity of these writings was questioned even in antiquity: see Linforth, *The Arts of Orpheus*, 168, 295, who argues that a good deal of Orphic literature may be a result of "the surprisingly common practice in Greece whereby poets sought to obtain prestige for their work by publishing it under the names of poets greater than themselves."

As to the prophetic powers of Orpheus, see Scholiast on Apollonius Rhodius, *The Voyage of Argo* 2.684; Ovid, *Metamorphoses* 11.8; Pliny the Elder, *Natural History* 7.203.

3. Orpheus is named by ancient sources as the institutor of rites of the Bacchic mysteries; see Apollodorus, *The Library* 1.3.2; Cicero, *On the Nature of the Gods* 3.58

For a discussion of the problems connected with Orpheus, see Harrison, *Prolegomena to the Study of Greek Religion*, chapters 9–12; Farnell, *Cults*, 5.105; W. K. C. Guthrie, *Orpheus and Greek Religion*;

Linforth, *The Arts of Orpheus*. Linforth is mainly concerned with the study of the religious tradition surrounding Orpheus. His main conclusion is that there was no such thing as an "Orphic religion" in antiquity. He suggests that the religious tradition associated with Orpheus arose from two independent sources: 1) the mystery religions of antiquity, and 2) the legend of Orpheus the magical singer.

Herodotus, *The Histories* 2.81, identifies the teachings of Orpheus with those of the Dionysiac religion.

4. J. E. Fontenrose, "Apollo and the Sun-God in Ovid," *American Journal of Philology* 61 (1940): 429–44. "Helius" is found as an epithet of Apollo in several late inscriptions of Asia Minor; for Patara, see *Journal of Hellenic Studies* 10 (1889): 81; for Thyateira, see *Corpus Inscriptionum Graecarum*, 10,3500. See also Plutarch, *On the E at Delphi* 86B, who scoffs at the identification of Apollo with the Sun-God. Harrison, *Prolegomena*, 462, mentions a Thracian cult of the Sun-God later fused with Apollo, and suggests that Orpheus may have been trying to revive the older Helius-cult when he came into conflict with Dionysus. Guthrie, *Orpheus and Greek Religion*, 46, notes that Dionysus was worshipped, and even identified, with Apollo at Delphi.

5. The descent of Orpheus to the underworld is recounted by Vergil, *Georgics* 4.469–503; Pausanias, *Guide to Greece* 10.30.6; Ovid, *Metamorphoses* 10.11; Apollodorus, *The Library* 1.3.2; Hyginus, *Fables* 164. See also Guthrie, *Orpheus and Greek Religion*, 29–32.

As to Eurydice, see Kern, *Orphicorum Fragmenta*, testimonia 62–67.

As to the violation of the tabu against looking back, see H. J. Rose, *Handbook of Greek Mythology*, 255; S. Thompson, *Motif-Index*, C331.2: "Travelers to other world must not look back"; F81.1: "Orpheus"; C331: "Person must remain in other world because of broken tabu."

6. As to the death of Orpheus, see Vergil, *Georgics* 4.521; Ovid, *Metamorphoses* 11.1; Pausanias, *Guide to Greece* 9.30.5–6. Various reasons are given for Orpheus's death: 1) he was slain by Zeus because he revealed certain mysteries to man; 2) he dared to go down to the underworld alive; 3) he offended Dionysus by forgetting to sacrifice to him; 4) he incurred the wrath of the Thracian women either by

excluding them from his mysteries or by avoiding their company; 5) Orpheus's mother offended Venus, who caused the Thracian women to become enamored of Orpheus and to dismember him. Orpheus was mentioned by some ancient sources as the founder of homosexual love (see Ovid, *Metamorphoses* 10.83). Leibethroe is usually named as the site of Orpheus's death. Pausanias, *Guide to Greece* 9.30.9 says his tomb was located there. Farnell, *Cults*, 5:105 discusses the cult of Orpheus at the site. The pieces of Orpheus's body were scattered, according to the traditional account, and his head, which retained the power of speech and prophecy, floated across the sea to Lesbos, where it became part of an oracle. The head continued to prophesy until Apollo ordered the oracle to be closed; see Guthrie, *Orpheus and Greek Religion*, 35.

7. As to the origin of this constellation, see Allen, *Star Names*, 280–88; Boll and Gundel, 904–6.

Ophiuchus

1. Hyginus is the earliest extant source for the stories of Carnabon, Triopas, and Phorbas. The name Triopas was applied to several Greek heroes, among them an old Thessalian hero, and a son of Helius. Hyginus's comment that Phorbas—clearly a Rhodian hero—was the son of Triopas and Myrmidon's daughter, Hiscilla, appears to conflate Thessalian and Rhodian tradition. The Myrmidons were inhabitants of Thessaly; Helius—along with Apollo, who is responsible for the changing of Phorbas into a constellation—was closely identified with the island of Rhodes.

For Triptolemus as the messenger of Ceres, see Apollodorus, *The Library* 1.5.2. On the cult of Triptolemus, see Farnell, *Cults*, 3:360.

2. Ancient sources generally refer to Asclepius as the son of Apollo and either Coronis or Arsinoe. Coronis seems the better choice, as she was recognized as Asclepius's mother by the priests of Asclepius at Epidaurus and by the oracle of Delphi. On the parentage and rearing of Asclepius by Chiron, see Apollodorus, *The Library* 3.10.3; Hyginus, *Fables* 202.

Athena gave Asclepius "the blood that flowed from the veins of the Gorgon." The blood that flowed from the veins on the left side had a baneful effect, while that from the right side was beneficial. Thus it was the latter which Asclepius used to raise the dead.

The story of a man learning to raise the dead from a snake is a common motif, occurring in Modern Greek, German, Italian and Lithuanian stories. See Frazer, *Adonis, Attis, Osiris*, 1:186; Thompson, *Motif-Index*, B512: "Medicine shown by an animal"; B511.1: "Snake as healer"; D965: "Magic plant"; E105: "Resuscitation by herbs"; E181: "Means of resuscitation learned."

For the references to Asclepius's appearance in the form of a snake, see G. Murray, *The Five Stages of Greek Religion*, chapter 1.

Lists of those men who were raised from the dead by Asclepius were compiled, among others, by Apollodorus, *The Library* 3.10.3.

3. Asclepius's death by lightning is mentioned as early as Hesiod (Fragment 125Rz). See also Apollodorus, *The Library* 2.10.3. Pindar, *Pythian Odes* 3.96, says that Asclepius was induced to raise Hippolytus from the dead by a large sum of money.

There were cults of Asclepius at Epidaurus, Athens, and in Thessaly. See Farnell, *Cults*, 4:239–41; W. Jayne, *The Healing Gods of Ancient Civilizations*, 250–52.

4. As suggested by Ps-Eratosthenes, who locates this constellation by referring to its brighter neighbor, Scorpio, Ophiuchus is not, itself, a bright constellation; none of its stars are of greater than the third magnitude, according to Ptolemy. See also Aratus, *Phaenomena* 79–80.

Orion

1. As to Orion's parentage, see Apollodorus, *The Library* 1.4.3.

2. The Greek word *hybris* connotes "insolence" or "wanton violence" arising from the pride of strength or from passion. *Nemesis* is sometimes personified.

Niobe boasted about her seven sons and seven daughters, provoking their death at the hands of Apollo and Artemis; Arachne boasted of her skill as a weaver and was changed into a spider by Athena; Tantalus and

Sisyphus were punished in the Underworld, the former for attempting to trick Zeus, the latter for attempting to cheat Death.

3. Greek heroes, having one mortal and one divine parent, were more than mortals, but less than gods. Many were founders of cities, and all were associated with a local cult. The best known Greek heroes include Heracles, Theseus, Perseus, Minos. See Farnell, *Greek Hero Cults*.

4. The cure of Orion's blindness by the Sun represents a folk-motif. See Thompson, *Motif-Index*, F952.2: "Blindness healed by rays of sun"; F950: "Marvelous Cures."

5. See p. 23.

6. See Boll and Gundel, 983–89; Allen, *Star Names*, 303–20.

Pegasus

1. Hippocrene on Mt. Helicon was created by Pegasus according to Strabo, *Geography* 8.6.21; Pausanias, *Guide to Greece* 9.31.3; Ovid, *Metamorphoses* 5.253.

As to the representation of this constellation in art, see Boll and Gundel, 928–29. *The Constellations* states that the constellation figure does not have wings. Aratus and Hipparchus do not mention anything at all concerning wings. It is interesting that in some of the earlier monuments, Pegasus was depicted without wings. See A. Baumeister, *Denkmäler des klassischen Altertums*, 1:301.

2. As to Mt. Helicon, see Strabo, *Geography* 9.2.25; Pausanias, *Guide to Greece* 10.28.1. Helicon and Cithaeron were once human brothers, who were changed by the gods into mountains. Helicon, being the more pious of the two, became the abode of the Muses, while Cithaeron became the home of the Erinyes. See [Plutarch], *On Rivers* 2; Scholiast on Homer, *Odyssey* 3.267. On Hippocrene, see Pausanias, *Guide to Greece* 9.31.3; Strabo, *Geography* 8.6.21. On the fountains at Troezen and Corinth, see Pausanias, *Guide to Greece* 2.31.9. On Aganippe, another famous fountain, see Pausanias, *Guide to Greece* 9.29.5. See also Leake, *Travels in Northern Greece*, 2:141, 205, 489–500, 526.

3. Hesiod, *Theogony* 278, 285; Apollodorus, *The Library* 2.3.2, 4.2; Scholiast on Aristophanes, *Peace* 722. On the slaying of Medusa by Perseus, see pp. 157–60.

See R. Engelmann, "Chimaira," in Roscher, *Lexikon*, 1,1:893–95.

4. See Apollodorus, *The Library* 1.9.3; Homer, *Iliad* 6.155; Hyginus, *Fables* 57; Pindar, *Olympian Odes* 13.63. See also S. D. Markman, *The Horse in Greek Art*, 8 and *passim*.

5. For additional references to Hippe see Scholiast on Germanicus BP 78.21, G 140.24. See also R. Wünsch, "Zu den Melanippen des Euripides," *Rheinisches Museum* 49 (1894): 94–96.

The Aeolus of this story is not the Greek god of the winds, but the son of Hellen and eponymous ancestor of the Aeolians.

6. As to the history of this constellation, see Boll and Gundel, 928–31; Allen, *Star Names*, 321–28.

Perseus

1. For the story of Danae and Perseus see Apollodorus, *The Library* 2.4.1; Pausanias, *Guide to Greece* 2.16.2, 25.6, 3.13.6; Hyginus, *Fables* 63; Herodotus, *Histories* 7.61. For the folk-motif of the floating chest see Thompson, *Motif-Index*, S141, 331: "Exposure in floating chest."

2. On Medusa, see Apollodorus, *The Library* 2.4.2. Aeschylus's *Phorcides* has not survived. Medusa was either a lovely maiden who was overly proud of her beautiful hair, which Athena, in punishment, changed into serpents, or she was changed into a monster by Athena for giving birth to Chrysaor and Pegasus in one of Athena's temples. In the usual account, however, Chrysaor and Pegasus (see pp. 151–55) were created by Poseidon from the blood of Medusa after she was slain by Perseus. See Hesiod, *Theogony* 287; Apollodorus, *The Library* 2.4.3; Ovid, *Metamorphoses* 4.792.

3. For the objects provided to Perseus by the various gods, see Hesiod, *Shield* 220; Apollodorus, *The Library* 2.4.3. Hades was the Greek god of the Underworld; Orcus, his Roman counterpart. Hyginus's insistence that Perseus used the helmet of Hades and not Orcus appears to be based on etymology: Hades probably means "the

invisible one," while the name Orcus may derive from the Greek word *horkos* meaning "oath." Hesiod mentions Orcus as a spirit that punishes perjury (*Theogony* 231; *Works and Days* 802).

On the Gorgon myth in connection with Athena, see Farnell, *Cults*, 1:286–88.

Lake Tritonis, into which Perseus threw the eye (and tooth) of the Graeae, thereby disabling them as guardians of the Gorgons, is located in Libya. See Ptolemy, *Geography* 4.3.19.

The rescue of Andromeda, which is not mentioned in connection with the constellation of Perseus in this story, was effected during Perseus's return journey to Seriphos. See pp. 27–28, 75–76, 83–84, 85. Before marrying Andromeda, Perseus defeated Phineus, Andromeda's uncle, to whom she had been betrothed. See Apollodorus, *The Library* 2.4.3. On Perseus after his return to Argos, see Ovid, *Metamorphoses* 5.236; Hyginus, *Fables* 244. Perseus was said to have founded Mycenae and Mideia by Pausanias, *Guide to Greece* 2.15.4. On the children of Perseus and Andromeda, see p. 28. Perseus was worshiped as a hero in the region between Argos and Mycenae. See Farnell, *Greek Hero Cults*, 337 and note.

4. On the history of this constellation, see Allen, *Star Names*, 329–35.

Pisces

1. As to the Great Fish, see pp. 163–65. The story of the two fish and the egg is told by Scholiast on Germanicus BP 81.20, G 145.9; Hyginus, *Fables* 197. Fish and doves were both sacred to the Dea Syria (Astarte). The story of Aphrodite and Eros related by Hyginus is also recounted by Manilius, *Astronomy* 4.579–81, 800–801; Ovid, *Fasti* 2.459–74 (Ovid says the two gods were only borne to safety on the backs of fish; they were not changed into fish). See also Boll and Gundel, 980–81, and J. G. Frazer, translator, *The Fasti of Ovid*, 2:390–93.

2. On the history of this constellation, see Boll and Gundel, 978–81; Allen, *Star Names*, 336–44.

Piscis Austrinus

1. As to the city of Bambyce/Hieropolis, see Strabo, *Geography* 16.1.27, 2.7; Pliny the Elder, *Natural History* 5.81, 32.17; Ptolemy, *Geography* 5.15.13, 8.20.8; Plutarch, *Life of Crassus* 17; Lucian, *On the Syrian Goddess* 1. The site is known today as Manbij. On the ruins of the ancient city, see D. G. Hogarth, "Hierapolis Syriae," *Annual of the British School at Athens* 14 (1907–08): 183–96; Cumont, *Etudes syriennes*, 22–24, 35. On the coinage of Hieropolis, see Head, *Historia Numorum*, 7.

2. As to the Eastern Mother Earth Goddess, see Ed. Meyer, "Astarte," in Roscher, *Lexikon*, 1,1:645–55; Cumont, "Astarte," in *Realenzyklopädie*, 2:1777–78; E. O. James, *The Cult of the Mother Goddess*. See also H. A. Strong and J. Garstang, *The Syrian Goddess*, 41. On the connection of Aphrodite with the Eastern Mother Earth Goddess, see Herodotus, *Histories* 1.105, 131, 199; Pausanias, *Guide to Greece* 1.14.7. See also Farnell, *Cults*, 2:626. As to the worship of the Syrian Goddess in Greece (Delos), see O. Rayet, "Dédicace à la déesse Atergatis," *Bulletin de Correspondence Hellenique* 3 (1879): 407. On the worship of the Syrian Goddess in the Roman world, see Apuleius, *The Golden Ass* 8.24–27.

3. As to the sacred fish of Derceto, see Lucian, *On the Syrian Goddess* 46; Pliny, *Natural History* 5.81. Lucian, *On the Syrian Goddess* 14, says that he saw an image of the goddess Derceto in Phoenicia which showed her as half-woman, half-fish, while at Bambyce, she was represented entirely as a woman. The Syrian tabu on fish is mentioned by Xenophon, *Anabasis* 1.4.9; Ovid, *Fasti* 2.473; Hyginus, *Fables* 197; Scholiast on Aratus, *Phaenomena* 386; Scholiast on Germanicus BP 98.16.

4. As to the history of this constellation, see Boll and Gundel, 1019–21; Allen, *Star Names*, 344–47. On possible reduplication in constellation figures, see Brown, *Primitive Constellations*, 2:220–22.

Planets

1. Cumont, "Les noms des planètes et l'astrolatrie chez les Grecs," *L'Antiquité Classique* 4 (1935): 5–43.

2. Homer, *Iliad* 23.226, 22.318. See also Hesiod, *Theogony* 381, 987; Pindar, *Isthmian Odes* 3.42.

3. As to the knowledge of the planets derived from the Babylonians, see Diodorus Siculus, *History* 2.30.3. Introduction of the names Phaenon, Phaethon, etc. is dated to the first half of the third century B.C.E. by W. Gundel and H. Gundel "Planeten," *Realenzyklopädie*, 20,2:2030. See also Cumont, "Les noms des planètes," 7–9; A. Florisoone, "Astres et constellations des Babyloniens," *Ciel et Terre* 67 (1951): 163–65. See also Pliny the Elder, *Natural History* 2.37; Ptolemy, *Tetrabiblos* 2.3.23. The planet Saturn was also called the star of Nemesis (see Achilles Tatius, in Maass, *Commentariorum in Aratum Reliquiae*, 43). The planet Mercury was also called the star of Apollo (see Pliny the Elder, *Natural History* 2.39).

The comparison in *The Constellations* of the reddish color of Mars to that "in the Eagle" is puzzling. Although the Greek is somewhat ambiguous, the reference is probably to the star α Aquilae (called "Eagle" in antiquity, see Ptolemy, *Almagest* 7.5), rather than to the constellation Aquila as whole. The comparison does not hold true today, however, as α Aquilae (Altair) is not a reddish star at present—its spectral type is A5 (white). It is interesting to note, also, that although Ptolemy designated this star as a second-magnitude star, its present magnitude is 0.89 and it is the twelfth brightest of all the stars (see Th. Page and L. W. Page, eds., *Starlight*, 93–94).

4. *The Constellations*, as well as the Scholiast on Germanicus BP, assign the descriptive names Phaenon and Phaethon to Jupiter and Saturn, respectively; Hyginus assigns the name Phaethon to both planets.

5. See p. 195. See also Scholiast on Germanicus BP 103.8. Compare Aratus, *Phaenomena* 10–11. The arrangement of the stars by Marduk is mentioned in the fifth tablet of the Babylonian creation epic; see J. B. Pritchard, ed., *Near Eastern Texts Relating to the Old Testament*, 67.

Pleiades

1. Hesiod, *Works and Days* 383–84, 619–23; Aratus, *Phaenomena* 265–67, 1085; Ovid, *Fasti* 3.105, 4.169–78; Vergil, *Georgics* 1.138.

2. The pursuit of the Pleiades by Orion is alluded to by Hesiod, *Works and Days* 619; Pindar, *Nemean Odes* 2.17; Scholiast on the *Iliad* 18.486 (the maidens are changed first into doves, then into stars). On the Pleiades' grief over their father, see Aeschylus, Fragment 312. On the Hyades, see pp. 191–92. On Orion, see pp. 147–50.

3. On the derivation of the name Pleiades, see Scholiast on Aratus, *Phaenomena* 254–55. Although the derivation of the name Pleiades from Pleione is possible, it was more usual for Greek offspring to be identified by their father's name; indeed, there are references to the Pleiades as *Atlantides*. All in all, given the numerous literary references to the Pleiades in their connection with sailing, that derivation of the name seems the most plausible.

4. Scholiast on Aratus, *Phaenomena* 254.

5. Allen, *Star Names*, 391–413.

Sagitta

1. Aeschylus, *Prometheus Bound*.

2. See Homer, *Iliad* 1.43–53, where Apollo rains his plague-infested arrows on the Greek camp. The arrows of Heracles, having been dipped in centaur's blood, were particularly venomous. The Scholiast on Germanicus G, 160.15 identifies the constellation with the spear used by Heracles to slay all the swans, a curious story if the text is not corrupt. However, it should be noted that Sagitta lies not far from the constellation Cygnus.

3. Homer, *Odyssey* 6.5, 9.106–15, 275, 383, 10.200; Hesiod, *Theogony* 139–41, 501–6; Apollodorus, *The Library* 1.1.2, 2.2.1.

4. On the slaying of Asclepius by Zeus, see pp. 141, 143. As to the Cyclopaean walls of Mycenae, see A. J. B. Wace, *Mycenae* 49–58; G. Mylonas, *Mycenae and the Mycenaean Age*, 19–21. On the slaying of the Cyclopes by Apollo, see Apollodorus, *The Library* 3.10.4.

5. On Apollo's exile and bondage, see Apollodorus, *The Library* 3.10.4. As to the laws pertaining to the punishment in cases of unpremeditated or justifiable homicide, see J. W. Jones, *Law and Legal Theory of the Greeks*, 251–76.

6. It is uncertain whether the Hyperboreans were a real or an imaginary race—the northern counterpart of Homer's "blameless Ethiopians" of the south. Some scholars argue that the Hyperboreans were Greek worshippers of Apollo in Thessaly or other northern parts of Greece (See Farnell, *Cults*, 4:100–106). As to the name of the Hyperboreans, see J. D. P. Bolton, *Aristeas of Proconnesus*, 195–96. Most scholars consider the Hyperboreans to be an imaginary people. Bolton contends that the information gathered by Aristeas concerning a people which he believed to be the Hyperboreans pertained in reality to the Chinese (Bolton, *Aristeas*, 100–101, 195). The earliest reference to the Hyperboreans is in the Homeric *Hymn to Dionysus* 29 (usually dated to the seventh or sixth century B.C.E.), where the name is synonymous with the end of the world. The lifestyle of the Hyperboreans was described by Pindar, *Olympian Odes* 3.28, 8.47; *Pythian Odes* 10.31–46; and also by Aristeas of Proconnesus in his lost epic, *Arimaspea*. Bolton surmises that the sole purpose of Aristeas's northward journey was to reach the land of the Hyperboreans. Aristeas did not reach his goal, but reported what he learned of the Hyperboreans from neighboring tribes. As to the suicide before senility of the Hyperboreans, see Pliny, *Natural History* 4.26, 89.

7. At Delphi, the return of Apollo from the Hyperboreans was celebrated in mid-summer. Apollo was believed to be among the Hyperboreans from the vernal equinox to the first setting of the Pleiades. The celebration of Apollo's return was called by other Greeks the "golden summer" of the Delphians. As to the Delian connection with the Hyperboreans, see Herodotus, *The Histories* 4.33. The bringing of the sacred gifts wrapped in wheat-straw at mid-summer bears an obvious resemblance to the "golden summer" of the Delphians. There was also a Delian tradition of Apollo's absence

among the Hyperboreans, although the scene was later changed to Lycia. Dodona was also believed to be a Hyperborean settlement. For a discussion of the role of Dodona in the transmission of the Hyperborean gifts to Delos, see H. N. Parke, *The Oracles of Zeus*, Appendix 3.

8. Pausanias, *Guide to Greece* 10.5.9.

9. Heraclides Ponticus wrote two dialogues in which Abaris was an interlocutor: *On Justice* and *Concerning Abaris*. The tradition that Abaris rode about on the arrow of Apollo may have begun with Heraclides. See Bolton, *Aristeas*, 158. As to Abaris himself, see Herodotus, *The Histories* 4.36. Abaris was said to have lived without food and to have saved Sparta from a plague (Pausanias, *Guide to Greece* 3.13.2).

10. Allen, *Star Names*, 349–51. Note that Ps-Eratosthenes and Hyginus appear to describe the figure of the arrow as facing in opposite directions. Each author speaks of four stars, but Ps-Eratosthenes mentions two stars on the notch, while Hyginus mentions two—presumably the same two—stars on the head of the arrow.

Sagittarius

1. As to the representation of centaurs in art, see E. Bethe, "Kentauren," in Roscher, *Lexikon*, 2,1:1074–80; Harrison, *Prolegomena*, 380; the "Battle of the Centaurs and the Lapithae" depicted on the metopes of the Parthenon (Elgin Marbles). In hunting scenes, the centaurs were portrayed with spears, not bows (see Xenophon, *On Hunting* 1.1). In battle scenes, the centaurs are armed, if at all, with rocks and branches (see Baumeister, *Denkmaler*, 1175, figure 1364; H. Oelschig, *De centauromachiae in arte graeca figuris*).

For the story of Ixion and Nephele, see Apollodorus, *Epitome* 1.20; Pindar, *Pythian Odes* 2.21–48; Hyginus, *Fables* 62. As to the centaurs in general, see G. Dumézil, *Le problème des Centaures*, 153–55; Rose, *Handbook of Greek Mythology*, 256.

As to the identification of the centaurs with the Modern Greek *kallikantzaroi*, see Lawson, *Modern Greek Folklore*, 190–255.

2. As to Chiron, see Apollodorus, *The Library* 1.2.4, 3.10.3; Homer, *Iliad* 11.832. Chiron was immortal but exchanged his immortality with Prometheus in order to end the pain he suffered from the poisoned arrow of Heracles (see Apollodorus, *The Library* 2.5.4). There was a cult of Chiron on the island of Thera (see Cook, *Zeus*, 1:142.)

3. As to the satyrs, see E. Kuhnert, "Satyros und Silenos," in Roscher, *Lexikon*, 4:444–531; Cook, *Zeus*, 1:696–99; Harrison, *Prolegomena*, 379–85. The satyrs were originally goat-like creatures. By the fifth century B.C.E., they were represented with horse-tails as a result of being confused with the horse-like sileni (older centaurs): see Cook, 1:696. According to Aristotle (in Lactantius Placidus on Statius, *Thebaid* 9.376), the satyrs were long-lived but not immortal.

4. As to Crotus, see Hyginus, *Fables* 224 (where he is called Croton), Scholiast on Germanicus BP 89.18, G 158.22. See also H. Gundel, "Krotus," in *Realenzyklopädie*, 11:2028–29. The Greek word *krotos* signified a rattling noise; it was used of the noise made by the feet in dancing and by the hands in applauding.

5. The evidence for the identification of Sagittarius with Centaurus is late: Lucan, *Pharsalia* 9.536; Ampelius, 2.9; Hermes Trismegistus, 67G. As to the identification of Sagittarius with Crotus, see Scholiast on Germanicus, BP 89.18, G 158.22. The figure of Sagittarius is not always represented as two-legged: Eudoxus, Aratus, and Ptolemy spoke of a four-legged creature.

As to the constellation of the Centaur, see pp. 79–82. As to the constellation of the Ship [Argo], see pp. 39–42. The constellation Argo rises when Sagittarius sets.

6. As to the history of Sagittarius, see Allen, *Star Names*, 351–60; Boll and Gundel, 967–71.

Scorpio and Libra

1. The earliest reference to Orion's death is in Homer, *Odyssey* 5.118. Other authorities who report that Orion died at the hands of Artemis include Callimachus, *Hymns* 3.264; Hyginus, *Fables* 195. Orion was stung by the scorpion because he tried to rape Artemis

(Scholiast on Homer, *Iliad* 18.486, *Odyssey* 5.121) or because of his boast that he could slay any animal on earth (Scholiast on Germanicus BP 63.7, G 122.6). Apollodorus, *The Library* 1.4.3, refers to other traditions concerning Orion's death: Orion was slain by Artemis either because he challenged the goddess to a match at quoits, or because he tried to rape one of the maidens in her train. The scene of Orion's death is located on Chios by the Scholiast on Aratus, *Phaenomena* 634; on Delos by Homer, *Odyssey* 5.123; on Crete by Hyginus.

2. Several ancient authors comment on the relative motion of the constellations Scorpio and Orion: Scholiast on Homer, *Iliad* 18.486; Aratus, *Phaenomena* 634. For other astral myths connected with Orion, see p. 150.

3. There was also an Egyptian constellation depicting a scorpion, but it may not have been identical with the Babylonian and Greek constellations. See Boll and Gundel, 966–67. It is likely that the Babylonian scorpion constellation was originally very large and included Ophiuchus and Sagittarius. The Claws of the scorpion are called *Zygos* ("balance") by Hipparchus. It was the Romans, however, who finally distinguished the Claws as a separate constellation to which they gave the name Libra ("balance"). The concept of a balance in connection with this constellation may be of Egyptian origin; see Allen, *Star Names*, 269–78, 360–72.

4. Allen, *Star Names*, 276–77.

Taurus

1. The "rape" of Europa is recounted, among others, by Ovid, *Metamorphoses* 3.847–75; Scholiast on Aratus, *Phaenomena* 167; Apollodorus, *The Library* 3.1.1; Hyginus, *Fables* 178; Nonnus, *Dionysiaca* 38.394. Europa was brought to Crete by the bull in most versions. Some late sources speak of her being hidden by Zeus on Mt. Teumessus near Thebes. The connection of Europa with Thebes is attested by the cult of Demeter Europa at Lebadeia, near Thebes, see Pausanias, *Guide to Greece* 9.39.5. As to the cult of Europa on Crete (where she was identified with the Eteocretan goddess of vegetation Hellotis), see

Willetts, *Cretan Cults and Festivals*, 152–68. Coins of the fifth century B.C.E. showing Europa riding a bull have been found at Gortyna and Phaestus on Crete. See Cook, *Zeus*, 1, figures 391–400. The motif is found on later coins from the Greek mainland, as well; see Willetts, *Cretan Cults*, 152–53.

2. Sir Arthur J. Evans, *The Palace of Minos*.

3. The constellation Taurus is identified with Io by Scholiast on Germanicus BP 74.20, G 135.18; Ovid, *Fasti* 5.619–20. The story of Io and Zeus is recounted by Aeschylus, *Suppliants* 291; Ovid, *Metamorphoses* 1.588, *Fasti* 4.717–20; Apollodorus, *The Library* 2.1.3; Lucian, *Dialogues of the Gods* 3. See also Herodotus, *The Histories* 1.1–2. On the connection of Hera with cows, see Cook, *Zeus*, 1:444–47, 453–57. See also Willetts, *Cretan Cults*, 111–12. Numerous votive cow-figurines have been excavated both at Argos (Heraeum) and at Olympia; see S. Eitrem, "Hera," in *Realenzyklopädie*, 8:369–403.

4. On the constellation as Pasiphae's bull, sometimes considered to be a separate animal from the bull of Marathon, see Scholiast on Germanicus G 136.1; Scholiast on Aratus, *Phaenomena* 167. See also Cook, *Zeus*, 1:464–67, 543–49. A late source (Scholiast on Germanicus G 96.11) says that the celestial bull is depicted as backing away from Orion (for whom see pp. 147–50), a story likely of astral origin.

5. On the genealogy and names of the Hyades, see R. Engelmann, "Hyades," in Roscher, *Lexikon*, 1,2:2752–58. On the Hyades as sisters of Hyas, see Homer, *Iliad* 18.486; Servius on Vergil, *Aeneid* 1.744. On the Pleiades, see pp. 171–73. On the Hyades as nurses of Dionysus, see Apollodorus, *The Library* 2.4.3; Hyginus, *Fables* 182, 192.

The name Hyades is derived from the shape of the constellation or from its connection with rain by Homer, *Iliad* 18.486; Scholiast on Euripides, *Ion* 1156; *Electra* 467; Vergil, *Aeneid* 1.744. The Roman name of this constellation, *Suculae*, is probably based on a misinterpretation of the Greek name Hyades, as deriving from the Greek word for "pig" (*hys*, *sys*), see Hyginus, *Fables* 192. The number of the Hyades varies between two and seven in ancient authors.

6. Ovid is uncertain whether the figure is a bull or a cow; see *Fasti*

4.717. On the history of this constellation see Boll and Gundel, 938–45; Allen, *Star Names*, 378–91.

7. Neither Ps-Eratosthenes nor Hyginus takes special note of the bright star Aldebaran (α Tauri).

Triangulum

1. The genitive form of the name Zeus in Greek (Διός) begins with the letter delta. Aratus describes this constellation as having the shape of an isosceles triangle (*Phaenomena* 235–36). The constellation is called *trigonon* ("triangle") by Eudoxus (in Hipparchus 1.2.13); Hipparchus 1.6.5; Ptolemy 7.5. It is called *deltoton* ("delta-shape") by Aratus, *Phaenomena* 235; Scholiast on Germanicus BP 81.7, G 144.

2. Scholiast on Aratus, *Phaenomena* 235; Scholiast on Germanicus BP 81.7, G 144.22. Hermes is mentioned again in *The Constellations* 43 as the arranger of the stars. Cf. Aratus, *Phaenomena* 10–11, who implies that the stars were arranged by Zeus.

3. Scholiast on Aratus, *Phaenomena* 233, "some say that the shape of the triangle among the stars is modeled on the situation of Egypt."

4. As to the history of this constellation, see Boll and Gundel, 933–34; Allen, *Star Names*, 414–16.

Ursa Major

1. For a detailed study of the story of Callisto, see R. Franz, "De Callistus Fabula."

The Arcadian origin of the story is attested by Callisto's connection with Pan, who is said by some to be her son; see Ar(i)aethus, Fragment 5; Scholiast on Theocritus, *Idylls* 1.123; A. Lang, *Myth, Ritual and Religion*, 2:181. Callisto is sometimes called Themisto or Megisto (Istrus, Fragment 57), but when she is referred to by either of these two names, Lycaon is not said to be her father. She is also called Helice and Phoenice (Servius on *Georgics* 1.246; Scholiast on Aratus, *Phaenomena* 27; Hyginus, *Fables* 177). Callisto is said to be the daughter of Ceteus or Nycteus by Apollodorus, *The Library* 3.8.2. In all other references she is said to be the daughter of Lycaon.

The "Aetolians" said by Hyginus to have captured Callisto constitute a curious intrusion into the story. The words "Aetolians" and "goatherds" are similar in Greek—*aetolon*, and *aepolon*, respectively— and the reading in Hyginus is probably, as Robert suggests (*Eratosthenis Catasterismorum Reliquiae*, 3), a scribal error. The same error occurs with reference to the constellation Bootes (see pp. 55–60).

As to the identification of Callisto with Artemis, see K. O. Müller, *Die Dorier*, 1,2:376; Lang, *Myth, Ritual and Religion*, 2:176–77, 208–20; Pausanias, *Guide to Greece* 8.3. See also V. Bérard, *De l'origine des cultes arcadiens*, 49–51; Farnell, *Cults*, 2:435–38.

Arcas is mentioned by several ancient authors as the son of Zeus and eponymous ancestor of the Arcadians: Pausanias, *Guide to Greece* 10.9.5; Scholiast on Apollonius Rhodius, *The Voyage of the Argo* 4.264.

2. As to Zeus Lycaeus and the sacred precinct on Mount Lycaeum, see Cook, *Zeus*, 2:63–70, 81–88; Farnell, *Cults*, 2:41–42; Pausanias, *Guide to Greece* 8.38.6; and Frazer, *Pausanias's Description of Greece*, 4:384.

The precinct of Zeus Lycaeus excavated on Mt. Lycaeum measures approximately 180 feet by 400 feet, and contains the bases of two columns described by Pausanias, *Guide to Greece* 8.38.6. The altar was covered by a layer of ashes five feet deep—apparently the remains of sacrifices. The bones among the ashes are mostly those of small animals. As to human sacrifice in the cult of Zeus Lycaeus, see Cook, *Zeus*, 1:70–81; and p. 200.

3. The story is first told by Hesiod (Fragment 181Rz). Amphis, who is the first to mention Zeus's appearance in the guise of Artemis, and Palaephatus (15) appear to draw on the Hesiodic version: see Franz, "De Callistus Fabula," 258. The version which comes after Hesiod but before Callimachus is attested by archaeological evidence and by Pausanias, *Guide to Greece* 8.3.6. Callimachus is followed by the Scholiast on Homer, *Iliad* 18.487 and Apollodorus, *The Library* 3.8.2. Callimachus was the first to mention that Callisto became a constellation, according to Franz, "De Callistus Fabula," 297. The constellation Ursa Major, however, was mentioned in Greek literature before

Callimachus (see Homer, *Iliad* 18.487, *Odyssey* 5.273), but was referred to as the "Wagon."

4. For a discussion of the various names of this constellation, see Allen, *Star Names*, 419–41. See also A. Scherer, *Gestirnnamen bei den indogermanischen Volkern*, 131–33, and pp. 201–4.

5. See Owen Gingerich, *The Great Copernicus Chase*, 10.

6. That the ancients thought of this constellation as consisting of seven stars is attested by Hipparchus, 1.5.6.

Ursa Minor

1. For the first tradition, see Aratus, *Phaenomena* 35; Servius, on *Georgics* 1.246, *Aeneid* 3.516; Scholiast on *Odyssey* 5.272. For the second tradition, see Servius on *Georgics* 1.246, 138. Ursa Major is always mentioned in connection with Ursa Minor in the first tradition, but not in the second. The names Cynosura ("dog's tail") and Helice are common to both traditions.

There are numerous literary references to Helice and Cynosura together as nurses of Zeus on Mount Ida: see Aratus *Phaenomena* 31–37; Scholiast on Homer, *Odyssey* 5.272; Servius on Vergil, *Georgics* 1.246, 138 (Servius refers to Helice and Cynosura as nymphs, but also calls Helice the daughter of Lycaon and relates about her the story of Callisto. He also identifies Cynosura with Phoenice, a nymph of Diana, who suffers the same fate as Callisto).

The city of Histoe was located on the south coast of Crete, between Hippocronium and Priansus (see E. B. James, "Crete," in Smith, *Dictionary of Greek and Roman Geography*).

Nicostratus was the son of Menelaus. His mother was either Helen or a slave woman: see Hesiod, Fragment 99Rz; Pausanias, *Guide to Greece* 2.18.6, 3.18.3, 3.19.9.

2. Thales is said to have discovered the constellation of Ursa Minor. See Scholiast on Homer, *Iliad* 18.487; Scholiast on Aratus, *Phaenomena* 27. Thales is reported to have urged the Greeks to follow the example of the Phoenicians in navigating by Ursa Minor rather than Ursa Major. In all probability, the name Phoenice, which Ps-Eratosthenes

assigns to a maiden whose fate is identical to that of Callisto, originally referred to the "Phoenician" constellation, and was later absorbed into Greek accounts of Ursa Minor.

3. See Franz, "De Callistus Fabula," 306–13, who believes that Ps-Eratosthenes drew his accounts of the two Bears and of Bootes from what was originally one story. Even so, the nature of Phoenice's "salvation" in the present story is not clear.

4. For the literary references to the two Bears as nymphs, wagons, or oxen, see H. Gundel, "Ursa, Sternbild," *Realenzyclopädie*, 9A,1:1051–54. See also Allen, *Star Names*, 447–49. The connection between the two constellations may have arisen from their similar configuration and their proximity in the sky.

Virgo

1. The myth of the Ages of Man is recounted by Hesiod, *Works and Days* 110–201, who tells of the disappearance of Aidos ("Shame") and Nemesis ("Divine Retribution") from the earth in the Iron Age. Aratus, along with Ps-Eratosthenes and Hyginus, imitates the passage in Hesiod, but substitutes Dike for Aidos and Nemesis. See also Ovid, *Metamorphoses* 1.89–149.

As to the parentage of the Horae, see Hesiod, *Theogony* 901–2.

2. In addition to the identifications provided by Ps-Eratosthenes and Hyginus, Virgo is identified with Thespia, eponym of the Boeotian city; Kore, the daughter of Demeter; Eileithyia, the goddess of childbirth; the Asiatic goddess Cybele; and the Greek goddesses Athena and Hecate. The name *parthenos* mentioned by Hyginus, was an epithet of Athena. As to the Babylonian constellation figure, see Allen, *Star Names*, 465.

For Erigone, see pp. 56–58.

Bibliography

A. Primary Sources

Achilles Tatius. In Maass, *Commentariorum*, 25–85.

Aelian. *On the Characteristics of Animals.* Translated by A. F. Scholfield. 3 vols. Loeb Classical Library. London: William Heinemann, 1958–59.

Aeschylus. In *The Complete Greek Tragedies.* Translated by D. Grene and R. Lattimore. Chicago: University of Chicago Press, 1969.

Aglaosthenes. In Jacoby, *Die Fragmente der Griechischen Historiker*, 3B:470–73.

Ampelius. *Liber memorialis.* Edited by E. Woelfflin. Lepizig: Teubner, 1873.

Amphis. In T. Kock, *Comicorum Atticorum Fragmenta*, 2:236–50. Leipzig: Teubner, 1884.

Anthologia Palatina. Edited by H. Stadtmüller. Leipzig: Teubner, 1894–1906.

Antoninus Liberalis. *Metamorphoses.* Edited by M. Papathomopoulos. Paris: Société d'Edition "Les Belles Lettres," 1968.

Apollodorus. *The Library.* Translated by Sir J. G. Frazer. 2 vols. Loeb Classical Library. London: William Heinemann, 1921.

Apollonius of Rhodes. *The Voyage of Argo.* Translated by E. V. Rieu. New York: Penguin Books, 1971.

Apuleius. *Metamorphoses.* Translated by J. A. Hanson. 2 vols. Loeb Classical Library. London: William Heinemann, 1989.

Aratus. *Phaenomena.* Translated by G. R. Mair. Loeb Classical Library. London: William Heinemann, 1976.

Ar(i)aethus. In Jacoby, *Die Fragmente der Griechischen Historiker*, 3B:26–30.

[Arion]. *Hymn to Poseidon*. In *Lyra Graeca*. Translated by J. M. Edmonds, 3:479. London: William Heinemann, 1980.

Aristeas of Proconnesus. In Bolton, *Aristeas of Proconnesus.*

Aristotle. *Generation of Animals*. Translated by A. L. Peck. Loeb Classical Library. London: William Heinemann, 1953.

————. *History of Animals*. Translated by A. L. Peck. 3 vols. Loeb Classical Library. London: William Heinemann, 1965, 1969.

————. *Meteorologica*. Translated by H. D. P. Lee. Loeb Classical Library. London: William Heinemann, 1952.

Arrian. *The Campaigns of Alexander*. Translated by P. A. Brunt. 2 vols. Loeb Classical Library. London: William Heinemann, 1983.

Athenaeus. *Banquet of the Sages*. Translated by C. B. Gulick. 7 vols. Loeb Classical Library. London: William Heinemann, 1927–41.

Augustine. *City of God*. Translated by G. E. McCracken et al. 7 vols. Loeb Classical Library. London: William Heinemann, 1957–78.

Bacchylides. *Bacchylidis carmina cum fragmentis*. 10th ed. Edited by H. Maehler. Leipzig: Teubner, 1970.

Boll, F. *Sphaera: neue Griechische Texte und Untersuchungen zur Geschichte der sternbilder*. Leipzig: Teubner, 1903.

Callimachus. Edited by R. Pfeiffer. 2 vols. Oxford: Clarendon Press, 1949–53.

Catalogus Codicum Astrologorum Graecorum. Edited by F. Cumont et al. Brussels: Lamertin, 1898–1936.

Catullus. *The Poetry of Catullus*. Translated by C. H. Sisson. New York: Orion Press, 1967.

Cicero. *On the Nature of the Gods*. Translated by H. Rackham. Loeb Classical Library. London: William Heinemann, 1979.

Claudian. *Battle of the Gods and the Giants*. Translated by M. Platnauer. 2 vols. Loeb Classical Library. London: William Heinemann, 1922.

Conon. In Jacoby, *Die Fragmente der Griechischen Historiker*, 1:190.

Diodorus Siculus. *Histories*. Translated by C. H. Oldfather. 12 vols. Loeb Classical Library. London: William Heinemann, 1968.

Eratosthenes. *Eratosthenis Catasterismorum Reliquiae*. Edited by C.

Robert. Berlin: Weidmann, 1963.

———. *Pseudo-Eratosthenis Catasterismi*. Edited by A. Olivieri. [*Mythographi Graeci* 3.1]. Leipzig: Teubner, 1897.

———. *Die geographischen Fragmente des Eratosthenes*. Edited by H. Berger. Leipzig: Teubner, 1880. Reprint. Amsterdam: Meridian, 1964.

———. *Eratosthenica*. Edited by G. Bernhardy. Berlin: Reimer, 1822. Reprint. Osnabrück: Biblio Verlag, 1968.

———. *Eratosthenis Catasterismi cum Interpretatione Latina et Commentario*. Edited by J. C. Schaubach. Göttingen, 1795.

Eudoxus. *Die Fragmente des Eudoxos von Knidos*. Edited by F. Lasserre. Berlin: De Gruyter, 1966.

Euripides. In *The Complete Greek Tragedies*. Translated by D. Grene and R. Lattimore. Chicago: University of Chicago Press, 1969.

Geographi Graeci Minores. Edited by K. Müller. Paris: Didot, 1855. Reprint. Hildesheim: Olms, 1965.

Hermes Trismegistus. In *Neue Astrologische Texte des Hermes Trismegistos*. Edited by W. Gundel. Munich: Akad. Munch., 1936.

Herodotus. *Histories*. Translated by A. De Selincourt. New York: Penguin Classics, 1954.

Hesiod. *Theogony, Works and Days, Shield*. Translated by Apostolos N. Athanassakis. Baltimore: The Johns Hopkins University Press, 1983.

———. *Fragmenta Hesiodea*. Edited by R. Merkelbach and M. L. West. Oxford: Clarendon Press, 1967.

Hipparchus. *In Arati et Eudoxi Phaenomena Commentariorum libri iii*. Edited by C. Manitius. Leipzig: Teubner, 1894.

Homer. *Iliad*. Translated by Richmond Lattimore. Chicago and London: Chicago University Press, 1967.

———. *Odyssey*. Translated by Richmond Lattimore. New York: Harper and Row, 1967.

Homeric Hymns. Translated by A. N. Athanassakis. Baltimore and London: The Johns Hopkins University Press, 1976.

Horace. *Odes, Epodes*. Translated by C. E. Bennett. Loeb Classical

Library. London: William Heinemann, 1927.

Hyginus. *Hyginus, De Astronomia*. Edited by Gh. Viré. Stuttgart and Leipzig: Teubner, 1992.

———. *Fabulae*. Edited by H. J. Rose. 3d ed. Leiden: A. W. Sijthoff, 1967.

———. *Poeticon Astronomicon*. Venice: Erhard Ratdott, 1482.

———. *The Myths of Hyginus*. Translated by M. Grant. Lawrence: University of Kansas Press, 1960. (Includes the *Fabulae* and Book 2 of the *Poetic Astronomy*.)

Istrus. In Jacoby, *Die Fragmente der Griechischen Historiker*, 3B:169–86.

Jacoby, F. *Die Fragmente der Griechischen Historiker*. Berlin: Weidmann, and Leiden: E. J. Brill, 1923–58.

Josephus. *The Jewish War*. Translated by H. St.J. Thackery. 2 vols. Loeb Classical Library. London: William Heinemann, 1927–28.

Kern, O., ed. *Orphicorum Fragmenta*. Berlin: Weidmann, 1922.

Livy. *History of Rome*. Translated by B. O. Foster. 14 vols. Loeb Classical Library. London: William Heinemann, 1967.

Lloyd-Jones, H., and P. Parsons. *Supplementum Hellenisticum*. Berlin: De Gruyter, 1983.

Longinus. *On the Sublime*. Translated by J. A. Arieti and J. M. Crossett. New York: E. Mellen Press, 1985.

Lucian. Translated by A. M. Harmon, K. Kilburn, and M. D. Macleod. 8 vols. Loeb Classical Library. London: William Heinemann, 1913–67.

Maass, E., ed. *Commentariorum in Aratum Reliquiae*. 2d ed. Berlin: Weidmann, 1898. Reprint, 1958.

Macrobius. Edited by F. Eyssenhardt. 2d ed. Leipzig: Teubner, 1893.

Manilius. Edited by A. E. Housman. 2d ed. 5 vols. Cambridge: University Press, 1937.

Martianus Capella. Edited by F. Eyssenhardt. Leipzig: Teubner, 1866.

Nicander. Edited and translated by A. S. F. Gow and A. F. Scholfield. Cambridge: University Press, 1953.

Nonnus. *Dionysiaca*. Translated by W. H. D. Rouse. 3 vols. Loeb

Classical Library. London: William Heinemann, 1940.

Numenius. In H. Lloyd-Jones and P. Parsons, *Supplementum Hellenisticum*, 279–85. Berlin: De Gruyter, 1983.

Oppian. *On Fishing*. Translated by A. W. Mair. Loeb Classical Library. London: William Heinemann, 1928.

Ovid. *Fasti*. Edited, with translation and commentary, by Sir J. G. Frazer. 5 vols. London: Macmillan, 1929.

———. *Metamorphoses*. Translated by F. J. Miller. 2 vols. Loeb Classical Library. London: William Heinemann, 1916.

Palaephatus. *De incredibilibus*. Edited by N. Festa. In *Mythographi Graeci* 3.2, 1–72. Leipzig: Teubner, 1902.

Pausanias. *Guide to Greece*. Translated by Peter Levi. Bungay: Richard Clay, 1971.

Pindar. Translated by Sir J. E. Sandys. Loeb Classical Library. London: William Heinemann, 1961.

Plato. *The Dialogues of Plato*. Translated by R. E. Allen. New Haven: Yale University Press, 1984.

Pliny the Elder. *Natural History*. Translated by H. Rackham and W. H. S. Jones. 11 vols. Loeb Classical Library. London: William Heinemann, 1940–63.

Plutarch. *The Parallel Lives*. Translated by B. Perrin. 11 vols. Loeb Classical Library. London: William Heinemann, 1914–26.

———. *Moralia*. Translated by F. C. Babbitt, W. C. Helmbold, P. H. DeLacy, et al. 15 vols. Loeb Classical Library. London: William Heinemann, 1927–69.

Powell, J. U. *Collectanea Alexandrina*. Oxford: Clarendon Press, 1925.

Ptolemy. *Almagest*. Edited by J. L. Heiberg. 2 vols. Leipzig: Teubner, 1898–1907.

———. *Ptolemy's Almagest*. Translated and annotated by G. J. Toomer. New York: Springer-Verlag, 1984.

———. *Geography*. Edited by F. A. Nobbe. Hildesheim: Georg Olms, 1966.

———. *Tetrabiblos*. Translated by F. E. Robbins. Loeb Classical Library. London: William Heinemann, 1940.

Quintus Smyrnaeus. Translated by A. S. Way. Loeb Classical Library. London: William Heinemann, 1913.

Scholia on Apollonius of Rhodes. Edited by C. Wendel. Berlin: Weidmann, 1958.

Scholia on Aratus. Edited by J. Martin. Stuttgart: Teubner, 1974.

Scholia on Aristophanes. Edited by W. Dindorf. Oxford: Clarendon Press, 1835–38.

Scholia on Callimachus. In Pfeiffer's edition.

Scholia on Euripides. Edited by W. Dindorf. 4 vols. Oxford: University Press, 1863.

Scholia on Germanicus. In Martianus Capella. Edited by F. Eyssenhardt. Leipzig: Teubner, 1866.

Scholia on Homer's *Iliad*. Edited by W. Dindorf. 6 vols. Oxford: Clarendon Press, 1874.

Scholia on Homer's *Odyssey*. Edited by W. Dindorf. Amsterdam: A. M. Hakkert, 1962.

Scholia on Pindar. Edited by A. B. Drachmann. Leipzig: Teubner, 1903, 1910.

Scholia on Theocritus. Edited by C. Wendel. Leipzig: Teubner, 1914.

Skylax. In Müller, *Geographi Graeci Minores*, vol. 1. Paris: Didot, 1855. Reprint. Hildesheim: Olms, 1965.

Strabo. *Geography*. Translated by H. L. Jones. 8 vols. Loeb Classical Library. London: William Heinemann, 1923–32.

Tacitus. *The Histories*. Translated by K. Wellesley. New York: Penguin Classics, 1964.

Theocritus. Edited by A. S. F. Gow. Oxford: Clarendon Press, 1952.

Valerius Flaccus. Translated by J. H. Mozley. Loeb Classical Library. London: William Heinemann, 1936.

Vergil. Translated by H. R. Fairclough. 2 vols. Loeb Classical Library. London: William Heinemann, 1934–35.

Xenophon. *Anabasis*. Translated by C. L. Brownson. Loeb Classical Library. London: William Heinemann, 1980.

B. Secondary Sources

Allen, R. H. *Star Names: Their Lore and Meaning.* New York: Dover, 1963.

Barnett, R. D. "Early Shipping in the Near East." *Antiquity* 32 (1958): 226.

Baumeister, A. *Denkmaler des klassischen Altertums zur Erlauterung des Lebens der Griechen und Römer in Religion, Kunst und Sitte.* 3 vols. Munich and Leipzig: R. Oldenbourg, 1885–88.

Beazley, J. D. *Attic Black-Figure Vase Paintings.* Oxford: Clarendon Press, 1956.

Bérard, V. *De l'origine des cultes arcadiens.* Paris: Thorin et fils, 1894.

Bethe, E., "Kentauren." In *Realenzyklopädie*, 11:172–78.

Boll, F., and H. Gundel. "Sternbilder." In Roscher, *Lexikon*, 6:901–1021.

Bolton, J. D. P. *Aristeas of Proconnesus.* Oxford: Clarendon Press, 1962.

Brown, Robert, Jr. *Researches into the Origin of the Primitive Constellations of the Greeks, Phoenicians and Babylonians.* 2 vols. Oxford: Williams and Norgate, 1899–1900.

Bruchmann, C. *Epitheta Deorum.* Supplement 3 to Roscher, *Lexikon*. Leipzig: Teubner, 1893.

Casson, Lionel. *The Ancient Mariners.* New York: Macmillan, 1959.

Chapouthier, F. *Les Dioscures au service d'une déesse.* Paris: E. De Bocard, 1935.

Cook, A. B. *Zeus: A Study in Ancient Religion.* 3 vols. New York: Biblo and Tannen, 1964.

Cumont, F. "Astarte." In *Realenzyklopädie*, 2:1777–78.

———. *Etudes syriennes.* Paris: Picard, 1917.

———. "Les noms des planètes et l'astrolatrie chez les Grecs." *L'Antiquité Classique* 4 (1935): 5–43.

Curtius, E. *Peloponnesus.* Gotha: Justus Perthes, 1851–52.

Demetrakou, D. *Mega Lexikon tes Hellenikes Glosses.* Athens: Archaios Ekdotikos Oikos Dem. Demetrakou, 1953.

de Molin, A. "De ara apud Graecos." Dissertation, Berlin, 1884.

de Navarro, J. M. "Prehistoric Routes between Northern Europe and Italy defined by the Amber Trade." *Geographical Journal* 66 (1925): 481–504.

Dilthey, J. F. K. "De electro et Eridano." Dissertation, Darmstadt, 1824.

Drexler, W. "Ganymedes." In Roscher, *Lexikon*, 1,2:1595–1603.

————. "Lykaon." In Roscher, *Lexikon*, 2,2:2168–73.

Dumézil, G. *Le problème des Centaures*. Paris: Librairie orientaliste Paul Genthner, 1929.

Düring, I. "Greek Music." *Journal of World History* 3 (1956): 307.

Easterling, P. E., and B. M. W. Knox, eds. *Greek Literature*. Cambridge: The University Press, 1985.

Eckels, R. P. *Greek Wolf-Lore*. Dissertation, University of Pennsylvania, Philadelphia, 1937.

Engelmann, R. "Chimaira." In Roscher, *Lexikon*, 1,1:893–95.

————. "Helena." In Roscher, *Lexikon*, 1,2:1928–78.

————. "Hyades." In Roscher, *Lexikon*, 1,2:2752–58.

Farnell, L. R. *Cults of the Greek States*. 5 vols. Oxford: Clarendon Press, 1896–1909.

————. *Greek Hero Cults and Ideas of Immortality*. Oxford: Clarendon Press, 1921.

Faulkner, R. O. "Egyptian Seagoing Ships." *Journal of Egyptian Archaeology* 26 (1940): 3–9.

Florisoone, A. "Astres et constellations des Babyloniens." *Ciel et Terre* 67 (1951): 153–69.

Fontenrose, J. E. "Apollo and the Sun-God in Ovid." *American Journal of Philology* 61 (1940): 429–44.

————. "Philemon, Lot, and Lycaon." *University of California Publications in Classical Philology* 13 (1950): 93–120.

Fowler, B. H. *The Hellenistic Aesthetic*. Madison: University of Wisconsin Press, 1989.

Frankfort, H. *Cylinder Seals: A Documentary Essay on the Art and Religion of the Ancient Near East*. London: Gregg Press, 1965.

Franz, R. "De Callistus Fabula." *Leipziger Studien* 12 (1890): 235–365.

Frazer, Sir J. G. *The Golden Bough*. 12 vols. New York: Macmillan, 1935.

———. *Pausanias' Description of Greece*. 5 vols. New York: Macmillan, 1898.

Friedländer, P. "Ganymedes." In *Realenzyklopädie*, 3:737–49.

Gingerich, Owen. *The Great Copernicus Chase*. Cambridge: Sky Publishing Corporation, 1992.

Gombosi, O. *Tonarten und Stimmungen der antiken Musik*. Copenhagen: E. Munksgaard, 1939.

Goold, G. P. "Perseus and Andromeda. A Myth from the Skies." *Proceedings of the African Classical Association* 2 (1959): 10–15.

Gundel, H. "Krotus." In *Realenzyklopädie*, 11:2028–29.

———. "Stephanos." In *Realenzyklopädie*, 3A,2:2352–61.

———. "Ursa, Sternbild." In *Realenzyklopädie*, 9A,1:1034–54.

———. "Prokyon." In *Realenzyklopädie*, 20:613–24.

———. and Gundel, W. "Planeten." In *Realenzyklopädie*, 20,2:2030–185.

Gundel, W. "Thyterion." In *Realenzyklopädie*, 6A:751–60.

Gürkoff, Emanuel. "Die Katasterismen des Eratosthenes." Dissertation, Würzburg, Sofia, 1931.

Guthrie, W. K. C. *Orpheus and Greek Religion*. 2d ed. London: Methuen, 1952.

Harris, J. R. *The Cult of the Heavenly Twins*. Cambridge: Cambridge University Press, 1906.

Harrison, J. E. *Prolegomena to the Study of Greek Religion*. London: Merlin Press, 1961.

Hartmann, A. "Schlange." In *Realenzyklopädie*, 2A,1:494–564.

———. "Silenos und Satyros." In *Realenzyklopädie*, 2A,3:35–53.

Head, B. V. *Historia Numorum*. Oxford: Clarendon Press, 1911.

Hogarth, D. G. "Hieropolis Syriae." *Annual of the British School at Athens* 14 (1907/8): 183–96.

Jaisle, K. *Die Dioskuren als Retter zur See bei Griechen und Romern*. Tubingen: Kommissionsverlag der J. J. Heckenhauerschen

Buchhandlung, 1907.

James, E. O. *The Cult of the Mother Goddess*. New York: Barnes and Noble, 1961.

Jayne, W. *The Healing Gods of Ancient Civilizations*. New Hyde Park, N.Y.: University Books, 1962.

Jones, J. W. *Law and Legal Theory of the Greeks*. Oxford: Clarendon Press, 1956.

Keller, O. *Antike Tierwelt*. Leipzig: W. Engelmann, 1913.

———. *Thiere des classischen Altertums*. Innsbruck: Wagner, 1887.

Kerenyi, C. *The Heroes of the Greeks*. Translated by H. J. Rose. London: Thames and Hudson, 1959.

Kirk, G. S. "Ships On Geometric Vases." *Annual of the British School at Athens* 44 (1949): 93–153.

Kuhnert, E. "Satyros und Silenos." In Roscher, *Lexikon*, 4:444–531.

Kunitzsch, Paul. *Der Almagest. Die Syntaxis Mathematica des Claudius Ptolelemäus in arabisch-lateinisher Überlieferung*. Wiesbaden: Otto Harrassowitz, 1974.

Lacey, W. K. *The Family in Ancient Greece*. London: Thames and Hudson, 1968.

Lang, A. *Myth, Ritual and Religion*. London: Longmans, Green, and Company, 1887.

Lawson, J. C. *Modern Greek Folklore and Ancient Greek Religion*. Cambridge: University Press, 1910.

Leake, W. *Travels in Northern Greece*. Amsterdam: A. M. Hakkert, 1967.

———. *Travels in the Morea*. Amsterdam: A. M. Hakkert, 1968.

Liddell, H. G., and R. A. Scott. *Greek-English Lexicon*. 9th ed. Oxford: Clarendon Press, 1940.

Linforth, I. M. *The Arts of Orpheus*. Berkeley and Los Angeles: University of California Press, 1941.

Lloyd, G. E. R. *Greek Science After Aristotle*. New York: W. W. Norton, 1973.

Lorentz, F. von. "Thyrsos." In *Realenzyklopädie*, 6A:747–52.

Maass, E. *Analecta Eratosthenica*. Vol. 6 of *Philologische Untersuchungen*.

Berlin: Weidmann, 1883.

Markman, S. D. *The Horse in Greek Art*. Baltimore: The Johns Hopkins Press, 1943.

Martin, J. *Histoire du texte des Phénomènes d'Aratos*. Paris: Librairie C. Klincksieck, 1956.

Meyer, Ed. "Astarte." In Roscher, *Lexikon*, 1,1:645–55.

Minto, A. *Il vaso François*. Florence: L. S. Olschki, 1920.

Morenz, S. "Die orientalische Herkunft der Perseus–Andromeda Sage." *Forschungen und Fortschritte* 36 (1962): 307–9.

Morrison, J. S., and R. T. Williams. *Greek Oared Ships*. Cambridge: Cambridge University Press, 1968.

Murray, G. *The Five Stages of Greek Religion*. Garden City, N.Y.: Doubleday, 1951.

Mylonas, G. *Mycenae and the Mycenaean Age*. Princeton: Princeton University Press, 1966.

Oelschig, H. "De centauromachiae in arte graeca figuris." Dissertation, Halle, 1911.

Otto, W. *Dionysos*. Frankfurt am Main: V. Klostermann, 1933.

Page, Th., and L. W. Page, eds. *Starlight*. Vol. 5 of Sky and Telescope Library of Astronomy. New York: Macmillan 1967.

Parke, H. W. *The Oracles of Zeus*. Oxford: Blackwell, 1967.

———. *Festivals of the Athenians*. Ithaca: Cornell University Press, 1979.

Pritchard, J. B., ed. *Near Eastern Texts Relating to the Old Testament*. 3d ed. Princeton: University Press, 1974.

Rayet, O. "Dédicace à la déesse Atergatis." *Bulletin de correspondance hellenique* 3 (1879): 407–10.

Reisch, E. "Altar." In *Realenzyklopädie*, 1:1640–91.

Roscher, W. H. *Ausführliches Lexikon der griechischen und römischen Mythologie*. 6 vols. and 4 supplements. Leipzig: Teubner, 1884–1937.

———. "Die Elemente des astronomischen Mythus vom Aigokeros." *Fleckeis. Jahrbücher* 151 (1895): 333–42.

———. "Die Sagen von der Geburt des Pan." *Philologus* 53 (1894):

362–77.

Rose, H. J. *Handbook of Greek Mythology.* New York: E. P. Dutton, 1959.

Russell, D. A. *Criticism in Antiquity.* London: Duckworth, 1981.

Sachs, C. *The Rise of Music in the Ancient World.* New York: Norton, 1943.

Sale, W. "The Popularity of Aratus." *Classical Journal* 61 (1965): 160–64.

———. "The Story of Callisto in Hesiod." *Rheinisches Museum* 105 (1962): 122–41.

Scherer, A. *Gestirnnamen bei den indogermanische Völkern.* Heidelberg: Carl Winter Universitätsverlag, 1953.

Seeliger, K. "Hesperiden." In Roscher, *Lexikon,* 1,2:2594–603.

Smith, W., ed. *Dictionary of Greek and Roman Antiquities.* 3d ed. 2 vols. London: J. Murray, 1890–91.

———. *Dictionary of Greek and Roman Biography and Mythology.* London: J. Murray, 1876.

———. *Dictionary of Greek and Roman Geography.* London: J. Murray, 1878.

Solmsen, F. "Eratosthenes' *Erigone*: A Reconstruction." *Transactions of the American Philological Association* 78 (1947): 252.

Stebbins, E. B. *The Dolphin in the Literature and Art of Greece and Rome.* Menasha, Wisc.: George Banta, 1929.

Steuding, H. "Theseus." In Roscher, *Lexikon,* 5:694.

Stoll, H. "Ariadne." In Roscher, *Lexikon,* 1,1:540–46.

Strong, H. A., and J. Garstang. *The Syrian Goddess.* London: Constable and Company, 1913.

Tarn, W. W. *Hellenistic Civilization.* 3d ed. Cleveland and New York: Meridian Books, 1966.

Theochares, P. A. "Iolcus." *Archaeology* 11 (1958): 13–18.

Thiele, G. *Antike Himmelsbilder.* Berlin: Weidmann, 1898.

Thompson, D'A. W. *A Glossary of Greek Birds.* Oxford: Clarendon Press, 1895.

Thompson, S. *Motif-Index of Folk-Literature.* Bloomington: Indiana

University Press, 1955–58.

Torr, C. *Ancient Ships.* Chicago: Argonaut, 1964.

Wace, A. J. B. *Mycenae.* Princeton: Princeton University Press, 1949.

Walcot, P. *Hesiod and the Near East.* Cardiff: Wales University Press, 1966.

Warmington, E. H. *Greek Geography.* New York: E. P. Dutton, 1934.

Webster, T. B. L. *Hellenistic Poetry and Art.* New York: Barnes and Noble, 1964

Wernicke, K. "Pan." In Roscher, *Lexikon,* 3,1:1347–481.

Willetts, R. F. *Cretan Cults and Festivals.* London: Routledge and Kegan Paul, 1962.

Williams, R. T. "Ships in Greek Vase-Painting." *Greece and Rome* 18 (1949): 126–37.

Winnington-Ingram, R. P. "Ancient Greek Music 1932–1957." *Lustrum* 3 (1958): 5–57.

Wissowa, G., ed. *Paulys Realenzyklopädie der Altertumswissenschaft.* Stuttgart: J. B. Metzler, 1894–1963.

Wünsch, R. "Zu den Melanippen des Euripides." *Rheinisches Museum* 49 (1894): 91–110.

Wycherley, R. E. *How the Greeks Built Cities.* London: Macmillan, 1962.

Yavis, C. G. *Greek Altars.* St. Louis: St. Louis University Press, 1949.

Author Index

Greek and Latin authors cited by Ps-Eratosthenes and Hyginus

General Index

144; associated with Sagitta, 175, 177; associated with Ursa Major, 199; *see also* Heracles

Hermes, 50, 109; aids Perseus 157, 159, 160; arranger of stars, 167, 195; places Lepus among constellations, 129; star of, 167, 169; *see also* Mercury

Herodotus, 47, 138, 179, 180

Hesiod, 17, 21–23, 38, 59, 106, 154, 172, 200, 207

Hesperus, *see* Venus, planet

Hipparchus, 25, 218 n.16, *passim*

Hippe, identified with Pegasus, 151, 152, 153

Homer, 16, 20–22, 23, 30, 35, 59, 67, 150, 168, 200

Horse, *see* Pegasus

Hyades, 21, 22, 24, 171, 173, 191, 192, 193, 259 n.5

Hydra, Crater, Corvus, 119–23

Hyperboreans, 175, 179–81

Iasion, identified with Gemini, 113

Icarius, 18; identified with Bootes, 56–60; associated with Canis Major, Canis Minor, 66, 67, 70; associated with Crater, 122

Io, identified with Taurus, 191, 192, 193

Isis, identified with Sirius, 65; rescued by Piscis Austrinus, 163; identified with Virgo, 205

Ixion, identified with constellation Hercules, 117

Juno, places Aquila among constellations, 34, 35; places Cancer among constellations, 62; places Draco among constellations, 101; Galaxy identified with milk of, 109; *see also* Venus, planet; Hera

Jupiter, places Capella among stars, 52; places Bootes among constellations, 56, 57; places Capricorn among constellations, 72; places Centaurus among constellations, 80; places Cygnus among constellations, 94; places Gemini among constellations, 112; places Hercules among constellations, 116, 117; places Leo among constellations, 126; places Ophiuchus among constellations, 142, 143; places Orion among constellations, 188; places Pegasus among constellations, 152; places Pleiades among constellations, 172; places Sagittarius among constellations, 184; places Scorpio among constellations, 188; places Taurus among constellations, 192; places Ursa Major among constellations, 56, 199; places Virgo among constellations, 57; planet, 167, 169; *see also* Zeus

Kids, *see* Haedi

Kneeler, *see* Hercules

Leo, 87, 125–28

Lepus, 23, 24, 69, 129–31

Liber, places Aries among constellations, 45, 46, 47; places Asini among constellations, 62; places Corona Borealis among constellations, 88, 89; places Delphinus among constellations, 98; *see also* Dionysus

Libra, 38, 187–89

Lion, *see* Leo

Lucifer, *see* Venus, planet

Lupus, 79, 80, 82, 232 n.6

Lycaon, identified with Lupus, 82

Lyra, 95, 116, 133–39

Maera (dog of Icarius), identified with Procyon, 57; identified with Canis Major, 66, 67
Maia (Pleiad), 109, 134, 171
Manger, see Praesepium
Marduk, 169
Mars, planet, 167, 168, 169; see also Ares
Medusa, 76, 152; beheaded by Perseus, 154, 157, 158, 159
Megisto, identified with Ursa Major, 199
Mercury, aids Perseus, 158; ordains movement of the stars, 168; places Aquila among constellations, 34; places Auriga among constellations, 51; places Lepus among constellations, 129; places Triangulum among constellations, 195; planet, 167, 168, 169; see also Hermes
Merope (invisible Pleiad), 171, 173
Merops, identified with Aquila, 34
Milky Way, see Galaxy
Minerva, associated with Erichthonius, 50–51; places Andromeda among constellations, 27; places Draco among constellations, 102; see also Athena
Muses, associated with Delphinus, 97, 100; associated with constellation Hercules, 116; place Lyra among constellations, 134, 135; ask Zeus to place Sagittarius among constellations, 184
Myrtilus, identified with Auriga, 50, 51

Nabu, 169
Neptune, places Delphinus among constellations, 98; see also Poseidon

Nereids, 76, 77, 231 n.3
Nergal, 169
Nicholas, Saint, 113
Nile River, identified with Eridanus, 105, 106; identified with Milky Way, 110; associated with Triangulum 195–96

Ocean, identified with Eridanus, 105, 106
Ophiuchus, 141–45, 248 n.4
Orion, 21, 22, 23, 24, 147–50; associated with Canis Major, 65, 66, 67; associated with Canis Minor, 69; associated with Lepus, 129; associated with Pleiades, 172, 173; associated with Scorpio, 187; associated with Taurus, 130
Orpheus, identified with constellation Hercules: 116, 117; Lyra as memorial to, 134
Orsilochus, identified with Auriga, 51, 53
Osiris, associated with Argo, 42

Pan, associated with Capricorn, 72–73
Parmenides, 110
Pausanias, 59, 138, 200
Pegasus, 41, 151–55, 249 n.1
Perigeios, 42, 105, 107
Perseus, 23, 24, 27, 84, 86, 122, 157–60
Phaenon, 167
Phaethon, 167
Phaon, identified with Gemini, 113
Philomelus, identified with Bootes, 58, 59, 60
Phoenice, identified with Ursa Minor, 201, 203–4
Pholus, identified with Centaurus, 80, 82

associated with Corona Borealis, 88; associated with Orion, 148; aids Perseus, 158; *see also* Hephaestus

Wagon, *see* Ursa Major
Water-Cup, *see* Crater
Water-Snake, *see* Hydra
Wild Animal, *see* Lupus

Zethus, identified with Gemini, 112
Zeus, abducts Ganymede, 29, 31, 33; associated with Aquila, 33, 35; associated with Ara, 37; associated with Triangulum, 195, 196; identified with Draco, 103; places Capella among stars, 50, 71; places Bootes among constellations, 55; places Canis Major among constellations, 65; places Capricorn among constellations, 71; places Centaurus among constellations, 79; places Cygnus among constellations, 93; places Gemini among constellations, 111; places Hercules among constellations, 101, 115; places Leo among constellations, 125; places Lyra among constellations, 134; places Ophiuchus among constellations, 141; places Orion among constellations, 147; places Pleiades among constellations, 172; places Sagittarius among constellations, 183; places Scorpio among constellations, 187; places Taurus among constellations, 191; places Ursa Major among constellations, 56, 197; places Ursa Minor among constellations, 203; places Virgo among constellations, 57; *see also* Jupiter

Visit the accompanying Star Myths Website at:

www.cosmopolis.com/star-myths